Microsoft Cybersecurity Architect Exam Ref SC-100

Second Edition

Ace the SC-100 exam and develop cutting-edge cybersecurity strategies

Dwayne Natwick

Graham Gold

Abu Zobayer

‹packt›

Microsoft Cybersecurity Architect Exam Ref SC-100

Second Edition

Authors: Dwayne Natwick, Graham Gold, and Abu Zobayer

Reviewers: Dan Gora and Jetro Wils

Publishing Product Manager: Anindya Sil

Development Editor: Richa Chauhan

Digital Editor: M Keerthi Nair

Presentation Designer: Shantanu Zagade

Editorial Board: Vijin Boricha, Megan Carlisle, Simon Cox, Ketan Giri, Saurabh Kadave, Alex Mazonowicz, Gandhali Raut, and Ankita Thakur

First Published: January 2023

Second Edition: October 2024

Production Reference: 1311024

Published by Packt Publishing Ltd.
Grosvenor House
11 St Paul's Square
Birmingham
B3 1RB

ISBN: 978-1-83620-851-8

www.packtpub.com

Contributors

About the Authors

Dwayne Natwick is the **CEO/Owner/Principal Trainer** at Captain Hyperscaler, LLC. He was previously the Global Principal Cloud Security Lead at Atos, a multi-cloud GSI. He has been in IT, security design, and architecture for over 30 years. His love of teaching led him to become an APMG-accredited ISACA trainer, a **Microsoft Certified Trainer** (**MCT**) Regional Lead and a **Microsoft Most Valuable Professional** (**MVP**), an AKYLADE Certified Instructor, and an ISC2 Authorized Instructor.

Dwayne has a master's degree in business IT from Walsh College; the CISM, CISA, and CRISC certifications from ISACA; the CISSP, CGRC, CSSLP, CCSP, SSCP, and CC certifications from ISC2; and over 18 Microsoft certifications, including Identity and Access Administrator, Azure Security Engineer, and Microsoft 365 Security Administrator. Dwayne can be found sharing information via social media, industry conferences, his blog site, and his YouTube channel.

Originally from Maryland, Dwayne currently resides in Michigan with his wife and three children.

To my wife, Kristy, thank you for always being there and supporting me. You are the love of my life and my best friend. To my children, Austin, Jenna, and Aidan, even with my career accomplishments, you are what makes me the proudest. You are all growing up to be such amazing people with kind hearts.

All four of you are my world and I could not make this journey without you. All my love and support for everything that you do.

– Dwayne Natwick

Graham Gold is a Senior Cloud Security Engineer at Admiral Group. He has 27 years' experience in financial services IT, now specializing in cloud security as of 2020. He has been instrumental in designing, building, securing, and running complex systems at enterprise scale across mainframes, Windows, Linux, and networks, on both on-premises systems and cloud platforms.

He is a multi-cloud certified professional, holding the Microsoft Cybersecurity Architect Expert, Azure Security Engineer, Google Certified Professional Cloud Security Engineer, and Google Certified Professional Cloud Architect certifications.

Graham is passionate about identity security and privileged access management, and loves to help his colleagues and community, sharing his knowledge on his blog and across social media platforms. Outside of work, he lives in Scotland with his wife and cats, and they share a love of world travel.

Abu Zobayer works as a Senior Cloud Solutions Architect at Microsoft, bringing over two decades of experience in the IT industry. Over the course of his career, he has taken on various key roles, such as Principal Microsoft Technical Trainer and Senior Customer Engineer. His credentials include a range of certifications: Microsoft Cybersecurity Architect Expert, Azure Security Engineer, Azure DevOps Expert, and Azure Solutions Architect Expert.

Abu holds a master's degree in cybersecurity from the University of Texas. He has played a crucial role in designing, deploying, and securing advanced cloud architectures, ensuring reliable and scalable solutions for enterprise-level clients.

Abu has a strong interest in cybersecurity and cloud innovations, and he frequently shares his expertise through training programs and community initiatives. Outside of his professional life, he enjoys experimenting with new technologies and spending quality time with his family in San Antonio, Texas.

About the Reviewers

Dan Gora is a Lead Cloud Security Architect at Eviden, part of ATOS, with over 15 years of experience in cybersecurity. Specializing in secure cloud transformation for highly regulated industries, he has guided organizations to enhance their security architecture by effectively implementing DevSecOps and zero-trust methodologies.

As an active contributor to the cybersecurity community, Dan is the OWASP Frankfurt Chapter Lead and Board Member of OWASP Germany. He has also co-authored several whitepapers for the Cloud Security Alliance. Dan holds a master's degree in secure software engineering from Darmstadt University of Applied Sciences, Germany, and certifications such as CISSP, CSSLP from ISC2, and CCSK from CSA, along with multiple credentials from Microsoft and AWS.

Originally from Germany, Dan now lives in Scotland with his civil partner, Margaretha.

To my partner, Margaretha, thank you for your unwavering love and support throughout the years. You are the cornerstone of my life and instrumental to my success. I cherish every moment with you.

– Dan Gora

Jetro Wils helps organizations operate safely in this cloud era by strengthening their information security and compliance, thus reducing risk and providing peace of mind. For 18 years, Jetro has been active in various tech companies in Belgium. Jetro's focus is practical cybersecurity advisory, specializing in cloud security, governance, compliance, and risk management. Jetro is a three-time Microsoft Certified Azure Expert and an MCT. He gives 10-20 certified training sessions annually on the cloud, AI, and security and has trained over 100 professionals, including enterprise architects, project managers, service managers, salespeople, team leaders, and engineers. He also hosts the BlueDragon Podcast, focusing on the above topics for decision-makers. Jetro is currently pursuing a master's degree in IT risk and cybersecurity management at the Antwerp Management School. He is a certified NIS 2 Lead Implementor, having gained the certification from PECB.

Table of Contents

2

Build an Overall Security Strategy and Architecture 37

3

Design a Security Operations Strategy 59

4

Design an Identity Security Strategy   87

5

Design a Regulatory Compliance Strategy   115

8

Design a Strategy for Securing SaaS, PaaS, and IaaS   191

9

Specify Security Requirements for Applications   221

10

Design a Strategy for Securing Data   237

11

Accessing the Online Practice Resources   261

Index   267

Other Books You May Enjoy   276

Preface

As the adoption of cloud infrastructure and services continues to grow at a rapid pace, cloud security has never been more critical. Businesses are increasingly moving their data, services, and applications to the cloud, creating a need for skilled professionals who can secure these environments. Cloud computing has evolved from a supplementary technology to a core competency within enterprises.

This shift has created a high demand for knowledgeable cloud security engineers and architects who can design, build, and operate secure cloud environments. The challenges posed by numerous security threats require organizations to develop robust cloud security strategies. Certifications play a vital role in identifying and developing the necessary skills for implementing cloud security measures. They also help individuals demonstrate their expertise to potential employers, advancing their careers.

The goal of this book is to equip you with the knowledge and skills needed to excel in cloud security. It covers a comprehensive range of topics essential for understanding and implementing cloud security measures. From cybersecurity fundamentals to advanced topics such as incident response, this book provides practical and straightforward explanations designed to educate you on the challenges and solutions in cloud security.

This book will prepare cybersecurity professionals like you for the SC-100 exam while also giving you a solid foundation that will help you put your knowledge to work and implement the strategies you learn. A mixture of theoretical and practical knowledge, practice questions, and a mock exam will ensure you breeze through the exam.

As you progress through this book, you will engage with various cloud security concepts and practices. The chapters cover critical areas such as cybersecurity in the cloud, building a security strategy, identity and access management, data protection, compliance, incident response, security operations, and future trends. Each chapter is designed to guide you through scenarios that test your understanding and application of cloud security principles.

By the end of this book, you will have a solid understanding of cloud security principles and practices and the confidence to apply this knowledge in your current role. You will be well prepared to tackle the challenges of securing cloud environments and stay ahead of emerging threats and technologies.

Who This Book Is For

This book is for a wide variety of cybersecurity professionals – from security engineers and cybersecurity architects to Microsoft 365 administrators, user and identity administrators, infrastructure administrators, cloud security engineers, and other IT professionals preparing to take the SC-100 exam. It is also a good resource for those who are designing cybersecurity architecture but not preparing for the exam. To get started, you will need a solid understanding of the fundamental services within Microsoft 365 and Azure, along with the security, compliance, and identity capabilities of Microsoft and hybrid architectures.

What This Book Covers

Chapter 1, Cybersecurity in the Cloud, provides an overview of cybersecurity and its evolution with cloud technologies. It explains how cybersecurity has changed as workloads have moved from on-premises data centers to the cloud.

Chapter 2, Build an Overall Security Strategy and Architecture, discusses developing and designing a security strategy for cloud, hybrid, and multi-tenant environments. It includes identifying integration points, translating business goals into security requirements, and designing security for resiliency.

Chapter 3, Design a Security Operations Strategy, covers designing and evaluating a strategy for security operations. Topics include logging and auditing for public, hybrid, and multi-cloud infrastructures, utilizing SIEM and SOAR solutions, and managing the incident life cycle.

Chapter 4, Design an Identity Security Strategy, focuses on creating an identity security strategy for cloud-native, hybrid, and multi-cloud environments. It emphasizes zero-trust principles and covers strategies for access management, conditional access, and privileged role access.

Chapter 5, Design a Regulatory Compliance Strategy, explores developing security and governance strategies based on regulatory compliance requirements. It includes using tools such as Microsoft Defender for Cloud and Azure Policy to evaluate and govern resources.

Chapter 6, Evaluate Security Posture and Recommend Technical Strategies to Manage Risk, discusses assessing security posture using benchmarks and tools such as Microsoft Defender for Cloud. It covers recommending security capabilities to mitigate identified risks.

Chapter 7, Design a Strategy for Securing Server and Client Endpoints, details creating security baselines and specifying security requirements for servers, mobile devices, and AD DS. It also covers managing secrets, keys, and certificates, and securing remote access.

Chapter 8, Design a Strategy for Securing SaaS, PaaS, and IaaS, involves building security baselines and specifying security requirements for various cloud services and workloads, including containers, edge computing, and application services.

Chapter 9, Specify Security Requirements for Applications, establishes security standards and strategies for applications and APIs. It includes prioritizing threat mitigation, onboarding new applications, and designing security solutions for API management.

Chapter 10, Design a Strategy for Securing Data, applies risk management frameworks and encryption standards to protect sensitive data. It covers identifying and protecting sensitive data and specifying encryption standards for data at rest and in motion.

How to Get the Most Out of This Book

This book is crafted to equip you with the knowledge and skills necessary to excel in the SC-100 exam through memorable explanations of major domain topics. It covers the core domains critical to cloud security and cybersecurity expertise that candidates must be proficient in to pass the exam. For each domain, you will work through content that reflects real-world cloud security challenges. At certain points in the book, you will assess your understanding by taking chapter-specific quizzes. This not only prepares you for the SC-100 exam but also allows you to dive deeper into a topic as needed based on your results.

Online Practice Resources

With this book, you will unlock unlimited access to our online exam-prep platform (*Figure 0.1*). This is your place to practice everything you learn in the book.

> **How to Access These Materials**
>
> To learn how to access the online resources, refer to *Chapter 11, Accessing the Online Practice Resources*, at the end of this book.

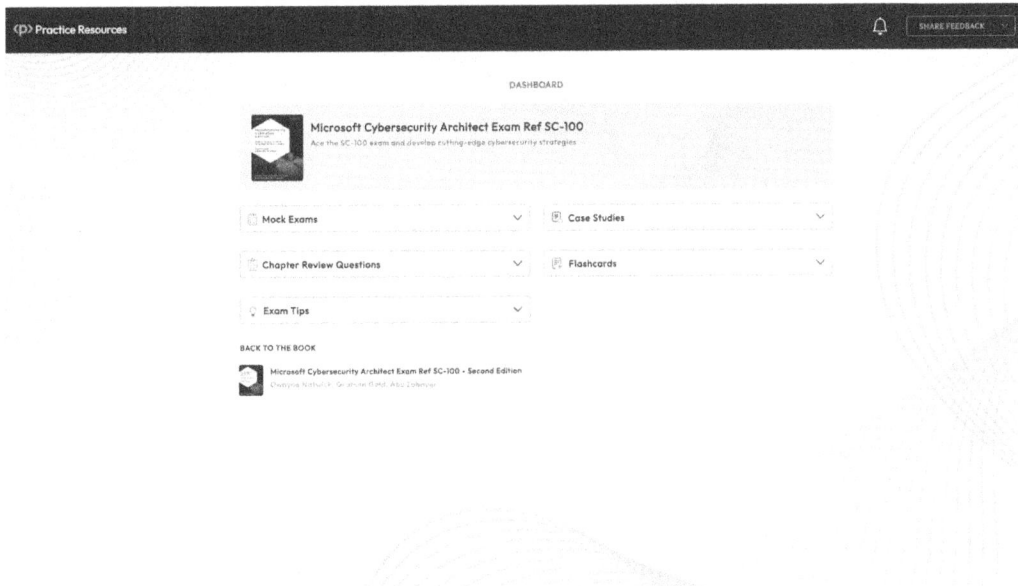

Figure 0.1: Online exam-prep platform on a desktop device

Sharpen your knowledge of SC-100 exam concepts with multiple sets of mock exams, interactive flashcards, case studies, and exam tips accessible from all modern web browsers.

Download the Color Images

We also provide a PDF file that has color images of the screenshots/diagrams used in this book. You can download it here: https://packt.link/SC-100_GraphicBundle.

Conventions Used

There are several text conventions used throughout this book.

Code in text: Indicates code words in text, database table names, folder names, filenames, file extensions, pathnames, dummy URLs, user input, and X (formerly Twitter) handles. Here is an example: "Since '1'='1' is always true, this query will always return all data from the users table, giving the malicious user access to all user accounts."

A block of code is set as follows:

```
SELECT * FROM users WHERE username = 'username' AND password =
'password'
```

Bold: Indicates a new term, an important word, or words that you see onscreen. For example, words in menus or dialog boxes appear in the text like this. Here is an example: "**Infrastructure as a Service (IaaS)** offers virtualized computing resources, including **Virtual Machines (VMs)**, storage, and networking. The user controls their infrastructure, while the **Cloud Service Provider (CSP)** oversees the physical hardware.

> **Tips or Important Notes**
> Appear like this.

Get in Touch

Feedback from our readers is always welcome.

General feedback: If you have any questions about this book, please mention the book title in the subject of your message and email us at customercare@packt.com.

Errata: Although we have taken every care to ensure the accuracy of our content, mistakes do happen. If you have found a mistake in this book, we would be grateful if you could report this to us. Please visit www.packtpub.com/support/errata and complete the form. We ensure that all valid errata are promptly updated in the GitHub repository at https://packt.link/SC100e2GitHub.

Piracy: If you come across any illegal copies of our works in any form on the internet, we would be grateful if you could provide us with the location address or website name. Please contact us at copyright@packt.com with a link to the material.

If you are interested in becoming an author: If there is a topic that you have expertise in and you are interested in either writing or contributing to a book, please visit authors.packtpub.com.

Share Your Thoughts

Once you've read *Microsoft Cybersecurity Architect Exam Ref SC-100, Second Edition*, we'd love to hear your thoughts! Scan the QR code below to go straight to the Amazon review page for this book and share your feedback:

https://packt.link/r/1836208510

Your review is important to us and the tech community and will help us make sure we're delivering excellent-quality content.

Download a Free PDF Copy of This Book

Thanks for purchasing this book!

Do you like to read on the go but are unable to carry your print books everywhere?

Is your eBook purchase not compatible with the device of your choice?

Don't worry – now, with every Packt book, you get a DRM-free PDF version of that book at no cost.

Read anywhere, any place, on any device. Search, copy, and paste code from your favorite technical books directly into your application.

The perks don't stop there: you can get exclusive access to discounts, newsletters, and great free content in your inbox daily.

Follow these simple steps to get the benefits:

1. Scan the QR code or visit the link below it:

https://packt.link/free-ebook/9781836208518

2. Submit your proof of purchase.
3. That's it! We'll send your free PDF and other benefits to your inbox directly.

1

Cybersecurity in the Cloud

This chapter will provide an overview of what cybersecurity is and why it matters in modern business.

It is important to look beyond news headlines and understand the business context, business challenges, threat scenarios, and impacts. Beyond passing the exam, the aim of this book is to enable you, as a cybersecurity practitioner, to protect your business while ensuring it can take advantage of business growth opportunities safely.

Often, you will discover that the choices that you need to make to balance these objectives are not binary choices; you need the business and threat context to make the correct decisions for your business. This chapter will also discuss the evolution of cybersecurity and cyber-attacks as cloud technologies have become more prevalent. Once you have completed this chapter, you will understand what cybersecurity means and how it has changed as we have moved our workloads from on-premises data centers to the cloud.

Overall, this chapter covers key exam domains and topics, specifically **Designing solutions that align with security best practices and priorities (20–25%)**. This includes creating a security strategy to support business resiliency, identifying and prioritizing threats to critical assets, and developing solutions for **business continuity and disaster recovery (BCDR)** in hybrid and multi-cloud environments, as well as mitigating ransomware attacks with a focus on BCDR and privileged access.

Making the Most of This Book – Your Certification and Beyond

This book and its accompanying online resources are designed to be a complete preparation tool for your **SC-100 exam**.

The book is written in a way that means you can apply everything you've learned here even after your certification. The online practice resources that come with this book (*Figure 1.1*) are designed to improve your test-taking skills. They are loaded with timed mock exams, chapter review questions, interactive flashcards, case studies, and exam tips to help you work on your exam readiness from now till your test day.

> **Before You Proceed**
>
> To learn how to access these resources, head over to *Chapter 11, Accessing the Online Practice Resources*, at the end of the book.

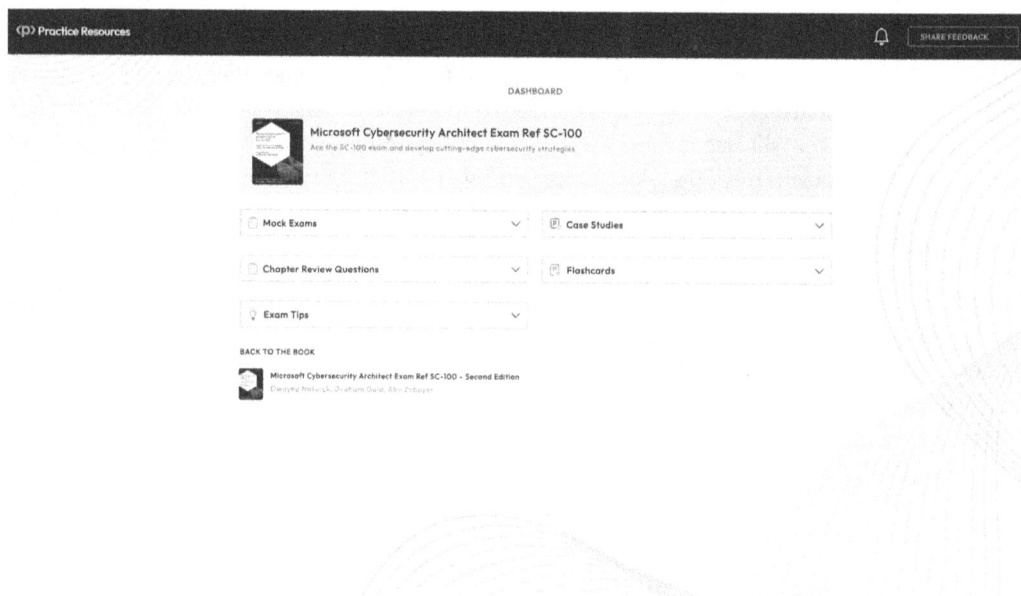

Figure 1.1: Dashboard interface of the online practice resources

Here are some tips on how to make the most of this book so that you can clear your certification and retain your knowledge beyond your exam:

1. Read each section thoroughly.

2. **Make ample notes**: You can use your favorite online note-taking tool or use a physical notebook. The free online resources also give you access to an online version of this book. Click the BACK TO THE BOOK link from the dashboard to access the book in **Packt Reader**. You can highlight specific sections of the book there.

3. **Chapter review questions**: At the end of this chapter, you'll find a link to review questions for this chapter. These are designed to test your knowledge of the chapter. Aim to score at least **75%** before moving on to the next chapter. You'll find detailed instructions on how to make the most of these questions at the end of this chapter in the *Exam Readiness Drill – Chapter Review Questions* section. That way, you're improving your exam-taking skills after each chapter, rather than at the end of the book.

4. **Flashcards**: After you've gone through the book and scored **75%** or more in each of the chapter review questions, start reviewing the online flashcards. They will help you memorize key concepts.

5. **Mock exams**: Revise by solving the mock exams that come with the book till your exam day. If you get some answers wrong, go back to the book and revisit the concepts you're weak in.

6. **Exam tips**: Review these from time to time to improve your exam readiness even further.

In this chapter, we are going to cover the following main topics:

- What is cybersecurity?
- The evolution of cybersecurity from on-premises to the cloud
- Cybersecurity architecture use cases
- Understanding the scope of cybersecurity in the cloud

What Is Cybersecurity?

To be able to understand the role of the **cybersecurity architect**, you should first understand what is meant by the term cybersecurity. The term is used in many different contexts within security, compliance, and identity.

Cybersecurity refers to the practice of protecting systems, networks, and programs from digital attacks. These cyber-attacks are usually aimed at accessing, changing, or destroying sensitive information, extorting money from users, or interrupting normal business processes.

Significance in Modern Business

In today's digital age, cybersecurity is crucial for several reasons:

- **Protection of data**: Businesses handle vast amounts of sensitive data, including personal information, financial records, and intellectual property. Cybersecurity measures help protect this data from breaches and theft.

- **Business continuity**: Cyber-attacks can disrupt business operations, leading to significant downtime and financial losses. Effective cybersecurity ensures that businesses can continue to operate smoothly.

- **Reputation management**: A data breach can severely damage a company's reputation. Strong cybersecurity practices help maintain customer trust and protect the brand's image.

- **Compliance**: Many industries are subject to regulations that require robust cybersecurity measures. Compliance with these regulations is essential to avoid legal penalties and maintain operational integrity.

Cybersecurity in the Context of the SC-100 Exam

The SC-100: Microsoft Cybersecurity Architect exam is designed for professionals who translate cybersecurity strategies into actionable capabilities that protect an organization's assets, business, and operations. Key areas covered in the exam include the following:

- **Zero-trust principles**: Implementing security strategies that assume breaches will occur and verifying each request as though it originates from an open network.

- **Identity and access management**: Ensuring that only authorized users have access to specific resources.

- **Platform protection**: Safeguarding the underlying infrastructure, including servers and networks.

- **Security operations**: Monitoring and responding to security incidents.

- **Data and AI security**: Protecting data and AI models from unauthorized access and manipulation.

- **Application security**: Ensuring that applications are secure from development through deployment.

- **Governance and risk compliance (GRC)**: Designing solutions that meet regulatory requirements and manage risk effectively.

Preparing for the SC-100 exam involves understanding these concepts and being able to design and implement security solutions that align with best practices and organizational needs.

To set a base level of understanding for this book, we will use the definitions provided by **NIST**, the **National Institute of Standards and Technology**. The reason for doing this is that many organizations use procedures and guidance from NIST and other agencies as the foundations of their own security standards, controls, and procedures.

According to NIST, there are multiple definitions for the term cybersecurity; the first part of the NIST definition is *"the prevention of damage to, protection of, and restoration of computers, electronic communications systems, electronic communications services, wire communication, and electronic communication, including information contained therein, to ensure its availability, integrity, authentications, confidentiality, and nonrepudiation."*

Cybersecurity is also defined by NIST as *"the prevention of damage to, unauthorized use of, exploitation of, and – if needed – the restoration of electronic information and communications systems and the information they contain, in order to strengthen the confidentiality, integrity, and availability of these systems."*

Taken together, this can be stated more simply: cybersecurity is the defense of electronic communications, systems, and information, ensuring that they remain available, accurate, and consistent, and confidential information remains so.

Notice also that there is emphasis placed on the ability to recover communications, systems, and information from any event, whether malicious or not.

Finally, notice that nonrepudiation is explicitly mentioned. It is not enough to be able to recover from an event; you must also be able to attribute every action or event to the true source of that event due to legal and regulatory obligations that most businesses will have to adhere to depending on their legal and geographic jurisdiction.

Overall, the underlying factors here are that you must take the steps to provide assurance for maintaining the confidentiality, integrity, and availability of your data and systems.

> **Note**
>
> The glossary at the following URL links to a plethora of NIST publications that give detailed cybersecurity guidance and, as such, is a notable example of how cybersecurity might be implemented in organizations that you work for now and in the future. It is advisable to read these documents. Though it is not required for the exam, they will be advantageous to you in your career in cybersecurity: `https://csrc.nist.gov/glossary/term/cybersecurity`.

In the next section, you will learn more about how the role of cybersecurity has changed from an on-premises to a cloud network and infrastructure.

Evolution of Cybersecurity from On-Premises to the Cloud

When protecting an **on-premises** data center and infrastructure, a cybersecurity architect designs various controls to safeguard physical assets and prevent unauthorized access at physical data center entry points or **internet service provider** (ISP) network entry points. Traditionally, these protections included a combination of physical security appliances, such as firewalls for packet inspection, and endpoint protection by allowing access to the data center only through SSL VPN-encrypted connections. These devices were managed by the company and given antivirus and anti-malware software to mitigate potential attacks.

As companies transition to more cloud-native applications, such as Microsoft 365, and build infrastructure on cloud providers like Microsoft Azure, the responsibility for security shifts from physical to virtual environments. This creates new vulnerabilities that the company must identify and plan ways in which to mitigate against threats. The following sections will discuss how a cybersecurity architect should plan for protection and controls within cloud and hybrid infrastructures.

Defense-in-Depth Security Strategy

When protecting cloud and hybrid infrastructure, there are many aspects that need to be considered. As you go through the various solutions offered within **Microsoft 365** and **Azure**, such as Microsoft Sentinel, the Microsoft Defender suite, and Microsoft Entra, the defense-in-depth methodologies and principles, which are explained in the next section, are essential for effectively protecting resources, identity, and data.

Building a Defense-in-Depth Security Posture

To protect your company from cyber-attacks, it is essential to implement controls that address each stage of an attack and maintain a defense-in-depth security posture. This approach ensures multiple layers of protection, making it harder for attackers to penetrate your defenses.

When considering your infrastructure, there are many logical layers that could potentially be breached through misconfiguration or exploitation of vulnerabilities.

These layers are shown in *Figure 1.2*.

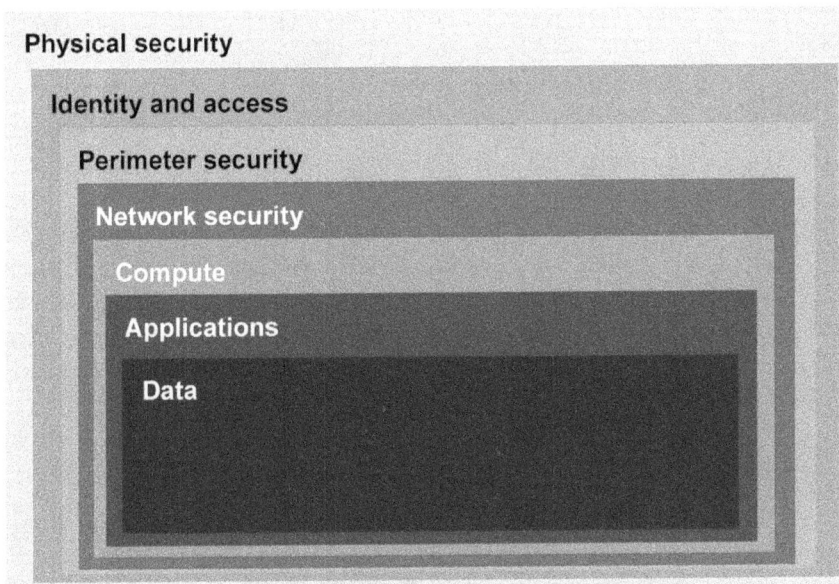

Figure 1.2: Logical layers of defense-in-depth posture/infrastructure

In *Figure 1.2*, we see the logical layers in the technology stack. It is these layers where an attacker may be able to gain access to systems and/or data.

In the following sections, you will explore these logical layers in depth, learning what each layer entails and how they can be secured.

Physical Layer

The **Physical** layer of defense includes the actual hardware technology and spans the entire data center facility. This includes the compute, storage, and networking components, rack spaces, power, internet, and cooling. It also includes the room that the equipment is housed in, the building location and its surroundings, and the processes that are in place for the guards, physical security staff, or guests that access these locations.

Protecting the physical layer encompasses how we create redundancy and resiliency in IT systems, and how we record and audit who accesses the building and systems. This could include gated fences, guard stations, video surveillance, logging visitors, and background checks. These physical controls should be in place for any company that utilizes its own private data center.

Although some of the considerations here may not seem related to an intrusion, it's important to remember that an attacker's goal is not always to access data. Sometimes, the objective is disruption, which is why redundancy and resiliency are mentioned in this section.

When utilizing Microsoft cloud services, the physical controls are Microsoft's responsibility. We will discuss shared responsibility for cloud security in the next section.

Identity and Access Layer

Since the provider is responsible for the physical controls within cloud services, **identity and access** become the first line of defense that a customer can configure and protect against threats. This is why statements such as *"Identity is the new control plane"* or *"Identity is the new perimeter"* have become popular when discussing cloud security. Even if your company maintains a private data center for the primary business applications, there is still a good chance that you are consuming a cloud application that uses your company identities or credentials. For this reason, having the proper controls in place, such as **multi-factor authentication (MFA)**, **conditional access policies**, and **Microsoft Entra Identity Protection**, will help to decrease vulnerabilities and recognize potential threats before a widespread attack can take place.

Perimeter Security Layer

Within a private data center, where the company controls the internet provider connection terminations and has firewall appliances, intrusion detection and protection solutions, and DDoS protection in place and fully configured, the protection of the perimeter is a straightforward architecture.

When working within cloud providers, **perimeter security** takes on a different focus. The cloud providers have agreements with the internet providers that provide services to their data centers, and these providers terminate these connections within their own hardware. The company perimeter security then becomes more of a virtual perimeter to their cloud tenant, rather than a physical perimeter to the data center network facilities. The company now relies on the provider's ability to protect against DDoS attacks at the internet perimeter.

Within Microsoft, DDoS protection is a free service since Microsoft wants to avoid a DDoS attack that would bring down many customers in a data center. For additional perimeter protection, the company can implement virtual firewall appliances to protect the tenant perimeter, to block port- and packet-level attacks, and additional solutions, such as Application Gateway, with a **web application firewall (WAF)** to protect from application-layer attacks.

Network Security Layer

The **perimeter** and **network security** layers work closely together. Both focus on the network traffic aspect of the company infrastructure. Where perimeter security handles the internet traffic that is entering the tenant, or data center, network security solutions protect how and where that traffic can be routed once it passes through the perimeter. Once an attacker can gain access to a system on the network, they will want to find ways to move laterally within the network infrastructure. Having proper IP address and network segmentation on the network can protect against this lateral movement taking place.

On a private data center network, this can be accomplished within switch ports with **virtual local area networks** (**VLANs**), configured to block traffic between network segments. In a cloud provider infrastructure, virtual networking, or VNETs, can accomplish similar network segmentation. In an Azure infrastructure, **network security groups** and **application security groups** can also be configured on network interfaces with additional port, IP address, or application-layer rules for how traffic can be routed within the network.

Compute Layer

After network security, we begin to get into the resources that hold our data. The first of these is our **compute** resources. To maintain clarity, we will generalize the compute layer as the devices with an operating system, such as Linux or Windows. Compute resources also include platform-based services where the compute layer is managed by the cloud provider, such as Azure App Service, Azure Functions, or containers. Within your own private data center with equipment that you own, protecting the host equipment and avoiding exposure by hardening the virtual hypervisor is necessary. In the public cloud, Microsoft or another cloud provider will be responsible for this. The customer responsibility for virtual machines in the cloud is focused on maintaining regular application of software updates and security fixes (often referred to as *patching*), to prevent exploitable vulnerabilities within the operating system. In addition, encrypting virtual machine operating systems and disks with Azure Disk Encryption will protect the disk images and contents from being exposed.

A common attack at the compute layer is scanning and gaining access to management ports on devices. Not exposing these ports, 3389 for Windows **Remote Desktop Protocol** (**RDP**) and 22 for the Linux **Secure Shell** (**SSH**) protocol, to the internet will provide a layer of protection against these attacks. Within Microsoft Azure, this can be accomplished with **network security group** rules, removing public IP addresses on virtual machines, **bastion hosts**, and/or utilizing **just-in-time virtual machine access**. Many of these security options will be discussed in *Chapter 7, Design a Strategy for Securing Server and Client Endpoints*.

Application Layer

The layer of defense that is closest to our data is our **applications**. Applications present data to users through our internet websites, intranet sites, and our line of business applications that are used to perform our day-to-day business. A cybersecurity architect will determine how to protect applications against common threats, such as cross-site scripting on our websites. To protect against these common threats, a WAF can be used for proper evaluation of the traffic accessing our applications. Using **Transport Layer Security** (**TLS**) protocol encryption can also help avoid the exposure of sensitive data to unauthorized individuals.

Prior to an application being moved to production, it should be rigorously tested to make sure that there are no open management ports and that all API connections are also secured.

If the application references connections to databases and storage accounts, the secrets and keys should not be exposed and a key management solution, such as **Azure Key Vault**, should be in place for the proper rotation of secrets, keys, and certificates. Properly securing these areas of our applications will help avoid exposure of sensitive data to those not authorized.

Data Layer

Always at the center of our defense-in-depth security posture is our **data**. Data is the primary asset of our company. This includes the business and financial data that is necessary for the company's survival and the personal information of our employees and customers. Exposure or theft of this information would have potentially catastrophic effects on the company's ability to continue. These effects could be reputational and involve financial loss.

As a security professional, one must protect data from intentional and accidental exposure to those who are not authorized to view it. Data resides in various areas within our technology infrastructure. Data can be found primarily in different storage accounts, such as blob containers or file shares, and within relational and non-relational databases. The widespread practice to accomplish this is through **encryption**.

Encryption makes data unreadable to those who are not properly authenticated and authorized to view it. Encryption can be used in diverse ways with data. First, there is encrypting data at rest, which is when it is stored and not being accessed. Next, there is encryption in transit, or while it is being delivered from where it is stored to the person requesting access. Finally, there is encryption in use, which maintains the encryption of the data within the application throughout the time that it is being viewed. This is the more complex of the types of data encryption since it requires the application to have the capability of presenting the encrypted data. Microsoft provides options for these encryption types that will be discussed later in this book.

Encrypting our data in our storage accounts and databases decreases its potential to be exposed to those not authorized. Additionally, requiring verification through authentication and authorization maintains the protection of data. This includes preventing anonymous access to storage accounts and masking sensitive data within our databases. The most important aspect of protecting our data is knowing where our sensitive data is located and planning proper steps to avoid it being exposed to the unauthorized. Bringing together the protection of data within the entire defense-in-depth strategy provides us with an effective way to protect against vulnerabilities and threats.

Maintaining a proper security posture across all defense-in-depth layers is the best way to protect our company from loss or exposure across cyber-attack stages. These stages will be further discussed later in this chapter. As security professionals, it is important that we take ownership of the planning, execution, monitoring, and management of all these layers and work with other stakeholders at each of these layers to maintain the overall security posture of the company.

Special considerations need to be accounted for within this security posture when utilizing public cloud services. In the next section, we will discuss how this shared responsibility for cloud services requires adjustments to our defense-in-depth security approach.

Shared Responsibility in Cloud Security

As technology has evolved and more resources have a level of exposure to external internet connections, the attack surface that is potentially vulnerable also increases. We must understand this and know where our responsibilities lie for each of the areas within our defense-in-depth security approach.

Remember that in the on-premises model, and in a hybrid cloud model, you may be solely responsible for everything within your data centers. This excludes the physical layer, which may be shared. You are likely not to build, own, and operate your data center facilities unless you are a large enterprise.

Where you operate in a cloud or hybrid model, the concept of shared responsibility comes to the fore and is the relationship between the customer and the cloud provider at each of the layers of defense in depth. This relationship differs depending on the technology that is being consumed.

Shared responsibility focuses on who has the ownership to interact at a specific level of protection. This may be physical ownership of equipment or administrative ownership for enabling various controls. The level of ownership between the company using the service and the cloud provider changes depending on the type of service that is being consumed by the company.

Table 1.1 shows shared responsibility for customers and Microsoft within the various cloud and on-premises services. We will learn more about these in depth later in the chapter; however, a brief description of them is as follows:

- **Infrastructure as a service (IaaS)**: IaaS provides virtualized computing resources over the internet. With Microsoft Azure, you can access essential infrastructure such as virtual machines, storage, and networking. This allows businesses to scale their IT resources as needed without investing in physical hardware.

- **Platform as a service (PaaS)**: PaaS delivers a platform that allows developers to build, deploy, and manage applications without worrying about the underlying infrastructure. Microsoft Azure App Service offers tools and services for application development, such as databases, middleware, and development frameworks.

- **Software as a service (SaaS):** SaaS provides access to software applications over the internet on a subscription basis. Microsoft Office 365 is a prime example, offering applications such as Word, Excel, and Outlook, which are hosted and maintained by Microsoft, eliminating the need for businesses to install and manage software on their own systems.

Responsibility	On-Premises	IaaS	PaaS	SaaS
Data governance and rights management	Customer	Customer	Customer	Customer
Client endpoints	Customer	Customer	Customer	Customer
Account and access management	Customer	Customer	Customer	Customer
Identity and directory infrastructure	Customer	Customer	Microsoft/ Customer	Microsoft/ Customer
Application	Customer	Customer	Microsoft/ Customer	Microsoft
Network controls	Customer	Customer	Microsoft/ Customer	Microsoft
Operating system	Customer	Customer	Microsoft	Microsoft
Physical hosts	Customer	Microsoft	Microsoft	Microsoft
Physical network	Customer	Microsoft	Microsoft	Microsoft
Physical data center	Customer	Microsoft	Microsoft	Microsoft

Table 1.1: Shared responsibility in the cloud

As you look at the customer's and Microsoft's responsibilities for security, the cybersecurity architect should determine the levels of controls that the company should have in place for each of the areas of potential vulnerabilities and exposure to attacks.

Understanding the Stages of a Cyber-Attack

Now that we have discussed the various layers within the defense-in-depth model, it is important to discuss the stages of a cyber-attack. This will demonstrate the importance of defense in depth. In other words, it will show why you cannot rely on a single security control in a single layer within the infrastructure.

There are many ways that an attacker can attempt to access resources within the company. How they gain this access and what they attempt to accomplish once they gain access is the foundation of a cyber-attack. *Figure 1.3* shows the stages of a cyber-attack in a linear format.

This is the Cyber Kill Chain, which is a cybersecurity model developed by Lockheed Martin in 2011. It outlines the stages of a cyber-attack to help security teams identify and stop malicious activities. The model is based on a military concept and breaks down an attack into several phases to aid in identifying and stopping attacker activity.

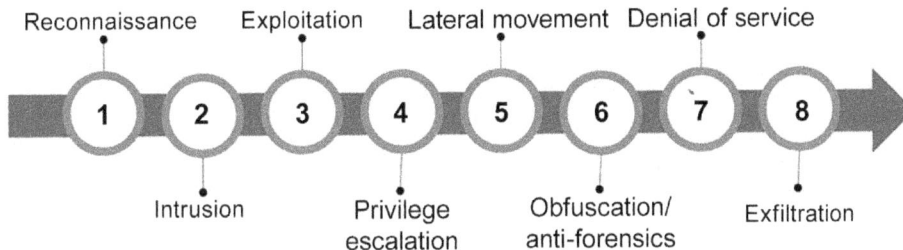

Reconnaissance   Exploitation   Lateral movement   Denial of service

1   2   3   4   5   6   7   8

Intrusion   Privilege   Obfuscation/   Exfiltration
                   escalation   anti-forensics

Figure 1.3: Stages of the Cyber Kill Chain

In many cases, an attacker attempts to enter and do some level of damage at one of these stages. Sophisticated attackers may go through every one of these stages to gain full access to resources and increase the amount of damage that they can do to a company. Let us define each of these stages for further understanding:

1. **Reconnaissance**: This is the planning stage of the attack, in which the attacker gathers information about the company or companies they will be targeting. They may use social media, websites, phishing, or social engineering of personnel within the company. Another aspect of this stage is port scanning of known management ports, such as RDP port 3389 or SSH port 22. The goal at this stage is to attempt to find ways to access systems. Port scanning helps determine which ports and services are open, closed, or filtered by sending packets to specific ports on a host and analyzing the responses.

2. **Intrusion**: Using the information gathered during reconnaissance, the attacker attempts to gain unauthorized access to the systems. One common method of intrusion is a brute-force attack, where the attacker tries multiple combinations of usernames and passwords to break into the system. This type of attack is described in more detail later in this chapter, in the *Common Threats and Attacks* section.

3. **Exploitation**: In this stage, the attacker has gained access to a system on the company network and now they want to exploit that system. A description of some of these attacks and exploits is covered in more detail later in this chapter, in the *Common Threats and Attacks* section. This is where the attacker begins to show malicious intent. They will begin to use this access to deliver malware across the network.

4. **Privilege Escalation**: Once the attacker has gained access to a system, they will want to elevate to administrator-level access to the current resource (privilege escalation), as well as additional resources on the network (lateral movement). If they have gained access to a virtual machine on the network, they could have administrative login privileges to other virtual machines and resources on the network.

5. **Lateral Movement**: Companies that use the same administrator username and password could allow the attacker to gain access to other systems across the network. This lateral movement could lead the attacker from a system without sensitive information to one that has extremely sensitive information.

6. **Obfuscation/Anti-Forensics**: As is the case with any attack or crime, the person or people involved do not want to be found or traced. Therefore, they attempt to keep their access anonymous. If they have gained access through someone's credentials within the company, this could help to decrease their traceability.

7. **Denial of Service**: When an attacker cuts off access to resources, this is a denial of service. This may be through an attack such as a SYN flood, where they send many requests to a company's public IP address that cannot be processed quickly enough. This flood of requests blocks legitimate requests from being able to access resources. Another means of denial of service could be a **ransomware** attack. This is not a typical blocking of information but more the withholding of information through encryption so that a company and its users can no longer access that information. The attacker then extorts the company for payment to make the information accessible.

8. **Exfiltration**: The final aspect of the cyber-attack is exfiltration. This is where the attacker gains access to sensitive information, and they can take that information to do harm in some way. This could be banking information, **personally identifiable information** (PII) about personnel or customers, or other valuable data.

The ability to protect against each of these aspects of the cyber-attack is how we break the Cyber Kill Chain/stop an attack before it can complete its ultimate objective. Each of these stages of the kill chain becomes an area to focus on protecting with cybersecurity controls. Understanding vulnerable areas and the potential threats to them in your infrastructure will allow you to determine ways to address and create a secure architecture.

Another popular framework is the **MITRE ATT&CK** framework. The MITRE ATT&CK framework was developed by the MITRE Corporation in 2013. It was initially created to document adversarial behavior and improve cybersecurity defenses by understanding how attackers operate.

Microsoft Defender for Cloud threat protection alert events are categorized based on the MITRE ATT&CK framework to understand and investigate potential attacks. *Figure 1.3* shows the Cyber Kill Chain.

For more information on the MITRE ATT&CK framework, visit `https://attack.mitre.org/`.

How Cybersecurity Architecture Can Protect Against These Threats

Now that we understand security posture, defense in depth, and shared responsibility, as you begin to architect cybersecurity for the cloud, we will discuss the makeup of a security operations team and the levels of a cybersecurity attack so that you gain an understanding of how cybersecurity architecture can help protect against these threats. You will see that a successful cybersecurity architecture is about more than just infrastructure and tooling; it is as much about people and processes as it is about technology.

Security Operations

In discussing security operations, you will hear terms such as red team, blue team, yellow team, purple team, white hat, and black hat. Let us define each of these:

- **Red team**: This is a team within the cybersecurity operation of the company that will conduct simulated attacks and penetration testing on the company infrastructure.

- **Blue team**: This team focuses on the defenses and the response to attacks. These are the incident responders within cybersecurity operations.

- **Yellow team**: These are developers and third-party developers that the blue team should be working with on defenses within the development of controls.

- **Purple team**: This team focuses on the methodology around the security architecture and protection. The purple team works closely with the red and blue teams to maximize the cybersecurity capabilities of the company. The purple team relies on the continuous feedback and lessons learned from the red and blue teams to improve the effectiveness of controls that are in place for vulnerability assessment, threat hunting and detection, and network monitoring.

- **White hat**: These are considered ethical hackers. Ethical hackers use the tools of a bad or malicious hacker to attack a company's systems but with their permission.

- **Black hat**: These are malicious hackers who are attempting to gain some level of control and do harm to the company that they are attacking.

Now that you understand the roles and responsibilities within the Security Operations department, the next section will discuss the scope of cybersecurity in a cloud infrastructure.

Understanding the Scope of Cybersecurity in the Cloud

A key to building a cybersecurity architecture is to know your responsibility as a cybersecurity architect and the responsibility of the cloud provider, depending on the type of services that you are utilizing.

In the following sections, you will learn how security controls will be utilized and put into place by the cybersecurity architect based on the shared responsibilities between the cybersecurity architect and providers.

Shared Responsibility Scope

It is important for a customer or company to understand their relationship to properly protect and secure their environment on the cloud. Let us discuss each of the services and the level of security responsibility. As a cybersecurity architect, you should think about how a control pertains to the shared responsibility model and to a defense-in-depth security approach.

On-Premises Responsibility

Although it may not seem directly relevant to a topic on cloud computing, most organizations that are not start-up businesses will commence their cloud journey with an on-premises infrastructure. They are likely to have a hybrid cloud infrastructure for many years after starting a cloud migration project, before finally having a fully cloud-native architecture (though for some businesses, it may not be practical or possible from an operational and/or regulatory perspective to be fully cloud-native).

On-premises infrastructure would be synonymous with a private data center. This is the equipment and infrastructure that the company owns. Therefore, the responsibility for security controls across all the levels of defense in depth is the company's responsibility. We have yet to consume any cloud services, so there is no responsibility for the cloud provider.

IaaS Shared Responsibility

Infrastructure-as-a-service, or IaaS, is the service that is most like a private data center. The primary difference between IaaS infrastructure and an on-premises data center is that the cloud provider is responsible for the physical security of the data center, any physical network equipment, and the hosts that provide our virtual servers. The customer is responsible for the following for IaaS:

- Putting all security controls in place to protect and patch the operating system
- Creating rules and infrastructure services such as firewalls to protect the network
- Managing and protecting applications from common threats
- Protecting identities and controlling access
- Patching and protecting endpoint devices

The customer is always responsible for the protection and governance of their data. This is shared across any of the cloud services in the shared responsibility model.

PaaS Shared Responsibility

Platform-as-a-service, or PaaS, removes the customer's responsibility for maintaining the operating system. The cloud provider handles all security patches and updates. Platform services have baseline security controls for the network, applications, and identity infrastructure. These are in place to protect against threats that could affect multiple customers who are utilizing these platform services. These baseline controls may not be seen as enough for some companies, so options to increase these controls are in place, and it is the customer's responsibility to turn them on. Many of these capabilities will be discussed later in this book. Within PaaS, the responsibility for access management, endpoint protection, and data protection and governance remains the sole responsibility of the customer.

SaaS Shared Responsibility

SaaS, or software-as-a-service, provides an application where you purchase a license on a per-user basis, log in to that application, and use it immediately. This simplifies these services to the consumer level, as there is a level of configuration that takes place for business applications. Microsoft 365 is an example of a SaaS application. The suite of software, Office 365 and SharePoint, for example, is available to use when you assign a license to a user. The cloud provider – in this case, Microsoft – has all the security controls in place for protecting the application, network, operating system, and physical environment.

Protection within SaaS is focused on identity and access management for the customer. Therefore, proper configuration of the identity and access controls is extremely important and ties into additional controls within endpoint protection, data protection, and governance. In a cloud infrastructure, SaaS, PaaS, and IaaS are all at play and need to be focused on within the cybersecurity architecture.

Note in *Table 1.1* that, although there may no longer be an on-premises infrastructure, there is a shared responsibility for the identity infrastructure. Microsoft does provide a level of security controls to protect user identities as a baseline, but the customer is responsible for increasing that level of protection. An example here would be turning on multi-factor authentication; it is provided by Microsoft, but the customer needs to enable the service for some or all users.

Many companies continue to have this private infrastructure while also utilizing public cloud services. These hybrid infrastructures vary across all the areas of responsibility to account for their overall security posture. As we continue through this book, the services that are discussed fall into one of the three main categories of IaaS, PaaS, or SaaS, but may also have a hybrid component to support on-premises infrastructure.

You now should have a strong understanding of defense-in-depth security and shared responsibility in the cloud. As you should have noticed, account and access management are an area of customer responsibility no matter what service is being consumed.

Principles of the Zero-Trust Methodology

In the previous section, we identified that the responsibility for securing the physical infrastructure for cloud services lies with the cloud provider, Microsoft. Since Microsoft is responsible for the first layer of defense in our defense-in-depth security posture, the first layer that we are responsible for as a company is the identity and access layer.

In *Chapter 2, Build an Overall Security Strategy and Architecture*, you will explore the role of identity and access management within a cloud and hybrid infrastructure and the services that Microsoft provides for protecting resources at this layer. It is important to understand the core concept that a company should adhere to when securing identity and access. This concept is the zero-trust methodology.

The **zero-trust methodology** is a process of continuously requiring someone on the network to verify that they are who they say that they are. The concept is straightforward and simple, but if you were to constantly ask users to enter their usernames and passwords, they would get frustrated.

To avoid this frustration, a zero-trust implementation utilizes various signals that alert about potentially anomalous behavior, leaked credentials, or insecure devices that trigger the need for a user to reverify their identity. These signals lead to a decision on what is needed to provide access to applications, files, or websites. This architectural pattern of zero-trust identity is shown in *Figure 1.4*:

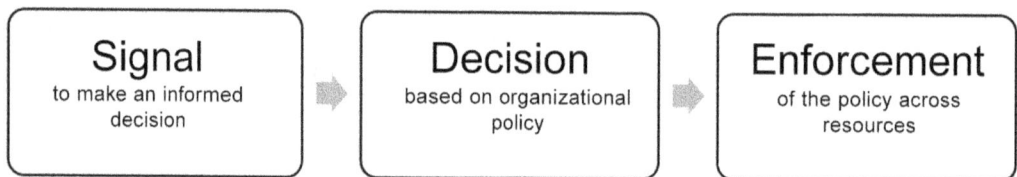

| Signal | Decision | Enforcement |
| to make an informed decision | based on organizational policy | of the policy across resources |

Figure 1.4: A flowchart where an initial signal leads to an informed decision based on organizational policy, which is then enforced across resources

> **Note**
>
> While not covered in detail in the exam, other important research to review on zero trust is the **Cybersecurity and Infrastructure Security Agency** (also known as **CISA**) **Zero Trust Maturity Model (ZTMM)**, recently updated to version 2.0. For more information on the CIA ZTMM, go to this link: `https://www.cisa.gov/zero-trust-maturity-model`.

As we discussed in the *Building a Defense-in-Depth Security Posture* section earlier in the chapter, in the defense-in-depth strategy, the physical controls are provided by Microsoft or the cloud provider; therefore, identity and access become the first layer of defense for a company and a cybersecurity architect to protect. The zero-trust model goes much further than simply identity and access, with networks, devices, applications, infrastructure, and data within the model and the defense-in-depth strategy. As demonstrated, there are several layers within your infrastructure where you could defend against attack, but also opportunities for attackers to gain access to your systems and data.

In the later *Defense in Depth: A Real-Life Example* section, we will demonstrate how this all comes together by looking at a real-world example of an attack.

A cybersecurity architect needs to know what the company can expect when it comes to vulnerabilities and attacks. The following sections will define some common internal and external threats and attacks.

Common Threats and Attacks

As cybersecurity architects, it is our responsibility to identify and design controls that address and protect against threats within our company infrastructure, whether on-premises, hybrid cloud, or cloud-native.

Threats can be internal or external. They also are not always malicious or meant to cause harm to the organization. We will discuss this in more detail as we identify some of these threats in the next sections. The threats listed are examples of internal and external threats and are not expected to be an exhaustive list.

When architecting a security operations infrastructure, many solutions utilize the MITRE ATT&CK framework for hunting for and identifying threats.

> **Note**
> For more information, please use the following link: `https://attack.mitre.org/matrices/enterprise/cloud/`.

This framework is extensive and covers the many diverse types of **tactics, techniques, and procedures (TTPs)** that can be used by attackers in combination, across several layers of the infrastructure, something that is often referred to as an **attack path**.

While the MITRE ATT&CK framework could fill a book, we can cover in this book some of the types of threats that are common, and later demonstrate how they can be chained together to form an attack path.

The **Cloud Security Alliance (CSA)** also provides guidance about common attacks and threats to cloud environments.

> **Note**
> More information can be found at this link: `https://cloudsecurityalliance.org/artifacts/security-guidance-v5`.

Finally, it is important to be aware of and understand the **Open Web Application Security Project (OWASP)** Top 10 Application Security Threats.

OWASP is a nonprofit organization dedicated to improving the security of software. It provides free and open-source tools, documentation, and training for web application security.

One of OWASP's most well-known projects is the OWASP Top Ten, a standard awareness document for developers on web application security. It highlights the most critical security risks to web applications and offers guidance on how to mitigate them.

The current Top 10 at the time of writing is as follows:

1. **Broken access control**: Improperly enforced restrictions on authenticated users, allowing them to access unauthorized functions or data.

2. **Cryptographic failures**: Issues related to the protection of data in transit and at rest, often due to weak or improperly implemented cryptographic algorithms.

3. **Injection**: Flaws such as SQL, NoSQL, and LDAP injection, where untrusted data is sent to an interpreter as part of a command or query.

4. **Insecure design**: Security weaknesses due to design flaws, rather than implementation issues.

5. **Security misconfiguration**: Incorrectly configured security settings or default configurations that are insecure.

6. **Vulnerable and outdated components**: Using components with known vulnerabilities or outdated software.

7. **Identification and authentication failures**: Issues with authentication mechanisms, such as weak passwords or flawed session management.

8. **Software and data integrity failures**: Problems with software updates, critical data, and CI/CD pipelines that can lead to unauthorized access or data corruption.

9. **Security logging and monitoring failures**: Inadequate logging and monitoring, which can delay the detection of breaches.

10. **Server-side request forgery (SSRF)**: When a web application fetches a remote resource without validating the user-supplied URL, leading to the potential exposure of internal systems.

These categories help organizations prioritize their security efforts and address the most pressing vulnerabilities.

> **Note**
> You can read more about OWASP at `https://www.owasp.org`.

Internal Threats

Internal threats are caused when a vulnerability is exposed by an internal user or resource. As stated previously, these are not always malicious or meant to cause harm; they can be accidental and created due to a lack of education and awareness. These internal threats, in some cases, can become vulnerabilities subject to external attacks. We will discuss this more as we discuss some of these internal threats in this section.

Shadow IT

Shadow IT is extremely common within companies. This is caused when people in the organization use applications not tested and approved by the company. Not all shadow IT causes a threat to the company, but not properly monitoring these applications can create vulnerabilities within the company. One way to discourage shadow IT is to have company policies in place regarding the use of third-party applications that are not approved on devices that access company resources. In addition, utilizing mobile device management or mobile application management can also deter the use of these applications by blocking access to them with device policies and conditional access. Educating users is another valuable aspect of stopping shadow IT from becoming prevalent within the company.

The life cycle of monitoring and preventing shadow IT within your company is shown in *Figure 1.5*:

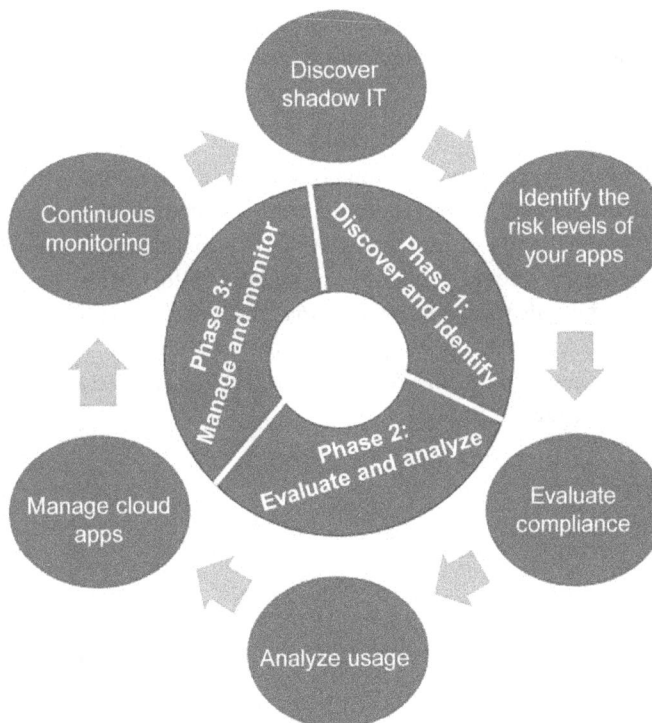

Figure 1.5: Shadow IT prevention life cycle

Figure 1.5 shows that **Phase 1** is **Discover and identify**, that is, identifying shadow IT (IT not managed by the IT department) and assessing the risk of those apps.

It then moves into **Phase 2**, where we evaluate whether the applications identified are compliant with relevant company and regulatory standards and risk appetite, before analyzing the usage of those applications.

Finally, we move into **Phase 3: Manage and monitor**. In this phase, we manage those discovered shadow applications, apply appropriate security controls, and then monitor them.

The entire process is depicted as a life cycle, as the security posture must be continuously assessed to ensure a secure environment.

Patch Vulnerabilities

Patch vulnerabilities are another internal threat to a company. These vulnerabilities can be created by users who defer patch installation and the restarting of their devices due to inconvenience. The most frequent patches that are provided for device operating systems are security patches. Therefore, if these patches are not installed company-wide in a timely manner, the entire company is vulnerable to potential exploitation. As was the case with shadow IT, a way to discourage deferring patch installation is through educating users on the risks that avoiding these updates poses to the company and their own devices. Automating patch updates and turning off the ability to defer them through mobile device management is also an option for companies to mitigate this threat.

Elevated Privileges

Elevated privileges are created when users have administrative rights to resources within the information technology environment that may not be required for them to complete the job tasks. A user with these privileges is an internal and external threat. As an external threat, if a user's credentials are compromised, then an attacker could gain access to sensitive information. As an internal threat, someone who has elevated privileges that allow them to access information that they are not required to view for their job could represent a privacy concern for the company. Therefore, it is important to review and audit user access and do our proper due diligence so that sensitive information is only available to those required to access it.

Developer Backdoors

When developing applications, access to the application infrastructure may be provided through an open port or service path. While the application is in development and isolated from the production infrastructure and data, this access helps developers gain access, work on, and test the application. However, if these developer backdoors are left in place after production, this could allow access to sensitive data and even access to application code that could be altered. Like privileged access, this could be thought of as an internal and an external threat. The exposure of these backdoors becomes a vulnerability that can be leveraged by attackers. It is an internal threat since it was created through the internal application development process.

Data Exposure

Data exposure is another threat here that could fall into both the internal and external threat categories. Companies must protect their sensitive data from being exposed to those not authorized to access it. Not having proper controls in place to protect sensitive data through access, authentication, and authorization could lead to exposure from either internal or external sources. Therefore, masking data from unauthorized users can protect against this exposure of data. Avoiding open and anonymous access to storage accounts will also protect against data exposure.

Perimeter Threats

The final internal threat that we will discuss in this section is perimeter threats. These threats are considered internal because they are created by inadequate controls in place to protect the internal infrastructure. Perimeter threats could be caused by allowing users to access resources through insecure open ports or transferring data through unencrypted transmission channels. IT professionals should have proper controls in place to avoid these threats and to monitor who is accessing data from inside and outside the company firewall.

As stated in the previous sections, internal threats can also become external vulnerabilities if not properly addressed with controls. It is an IT professional's responsibility to use proper due care and due diligence to protect the company.

Now that we have discussed some potential internal threats, let us review some potential external threats.

External Threats

The previous section focused on threats that are created internally by users, developers, or IT staff that could cause data exposure to unauthorized personnel or allow external attackers into the company infrastructure. In this section, we will discuss external threats that are initiated by external sources. These external threats can cause disruption to the company and customers, causing decreases in efficiency and revenue.

Denial-of-Service Attacks – Network and Application Layer

Denial-of-service attacks are a common external threat to companies. Also referred to as **distributed denial-of-service**, or **DDoS**, these attacks flood your ISP with thousands of requests to overwhelm the ISP and the company infrastructure to the point that actual users attempting to access resources cannot get through and their requests time out. A DDoS attack is not a threat that is based on theft, and no personal or company data is at risk during these types of attacks. These attacks are damaging to a company from a revenue and efficiency standpoint. For example, remote internal users may not be able to access the resources required to perform their job-related tasks. In addition, customers may not be able to access the company website to browse and order, costing the company revenue.

Figure 1.6 shows how these attacks threaten the ability of an actual user to access a system:

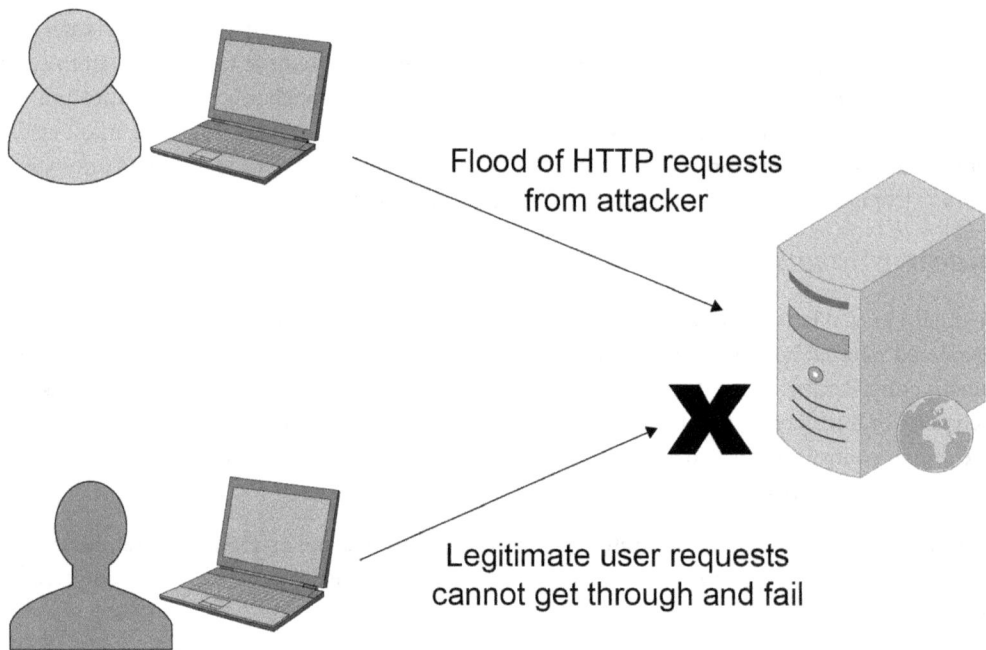

Figure 1.6: Illustration of a denial-of-service attack

The longer that a company is subject to these types of attacks, the greater the cost in lost revenue and time. Therefore, it is important that a company monitors these attacks and can block their source quickly to minimize the impact.

Brute-Force Attacks – Network and Application Layer

In contrast to a DDoS attack, where there is no threat of personal or company data being stolen, this is not the case with a brute-force attack. A brute-force attack is a threat with the primary purpose of gaining access to a company's systems to digitally burglarize data. Brute-force attack threats are commonly tied to some of the internal threats mentioned previously in this chapter. These types of threats attempt to gain access to the company systems by finding an opening within those systems and then, as the name suggests, using brute force to access them. These attacks are carried out by scanning for ports that are open to the internet, finding systems that have public internet addresses on those ports, and then using commonly used usernames and passwords on systems to gain access.

Figure 1.7 shows how an attacker utilizes multiple systems and attempts to gain access to systems:

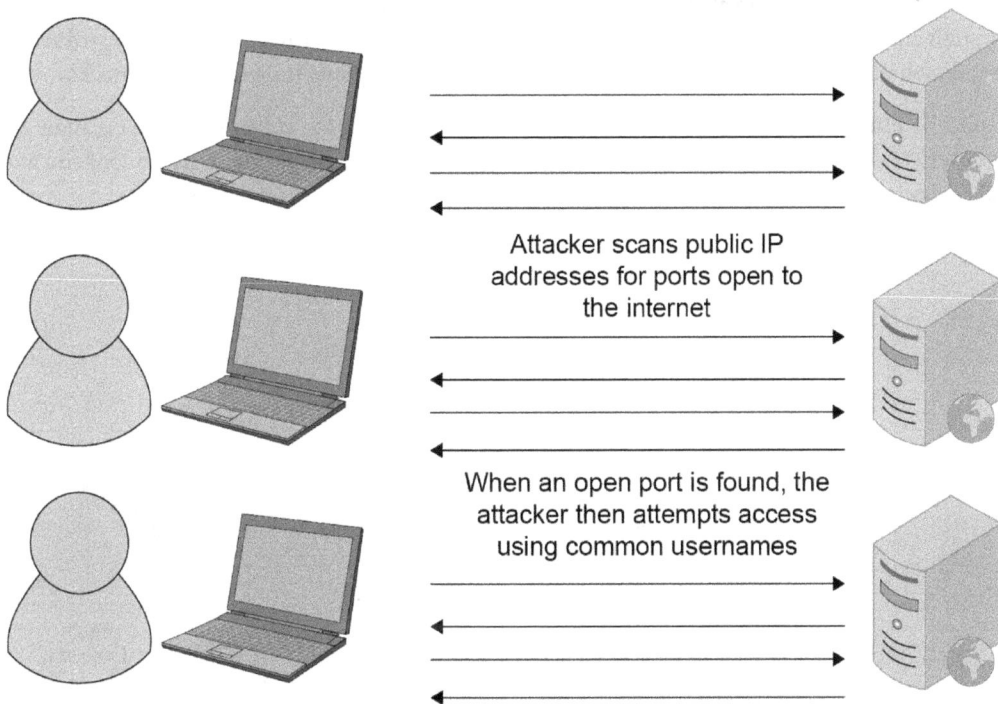

Attacker scans public IP addresses for ports open to the internet

When an open port is found, the attacker then attempts access using common usernames

Figure 1.7: An attacker scanning public IP addresses for open ports, then attempting
to gain access using common usernames once an open port is found

When a brute-force attack is successful, the company is exposed to potential theft of sensitive personal or company data that may be on that system, or other databases and file shares that are accessible from that system.

Software Vulnerabilities – Application, Network, Endpoint, Identity, and Access Management Layers

Software vulnerabilities allow external threats where attackers take advantage of some of the controls that are not in place to protect the company. Some of these vulnerabilities can be caused by the internal threats that were mentioned in the previous section, such as development backdoors and patch vulnerabilities. Improperly securing application APIs also creates a vulnerability that an attacker can exploit. The threat of a software vulnerability may lead to data breaches where an attacker can gain unauthorized access to sensitive information and applications.

Many vulnerability exploits are caused by operating system code, third-party libraries, or application code that an attacker has found could be exploited. These are called zero-day exploits and are the most common of widespread threats to systems. Keep in mind that this is an external attack but is initiated through an internal user accessing a malicious email or link. Proper user education regarding the origination of emails and links can assist in preventing these exploits from becoming attacks.

Figure 1.8 illustrates the life cycle of a zero-day threat, detailing the stages from the creation and discovery of a vulnerability, through the availability of an exploit, the period of risk before and after public disclosure, and finally, the release and installation of a patch by the vendor.

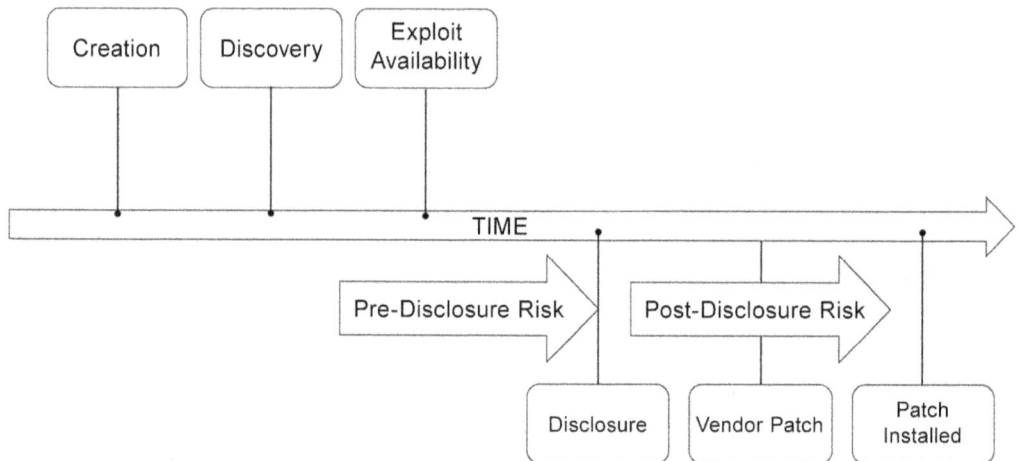

Figure 1.8: The vulnerability management life cycle

IP or Identity Spoofing

An IP or identity spoofing threat comes from an attacker pretending to be someone within the company or utilizing an IP address that is seen by systems as internal. Attackers that leverage these threats have gathered information on the company through some type of phishing campaign that has allowed them to identify usernames, passwords, and IP addresses that have access to systems. These attacks are used to gain access to systems. Social engineering and phishing attacks are methods that can be used to gain this level of access.

Figure 1.9 shows an attacker that has gained access to an authorized user's identity to gain access to another user:

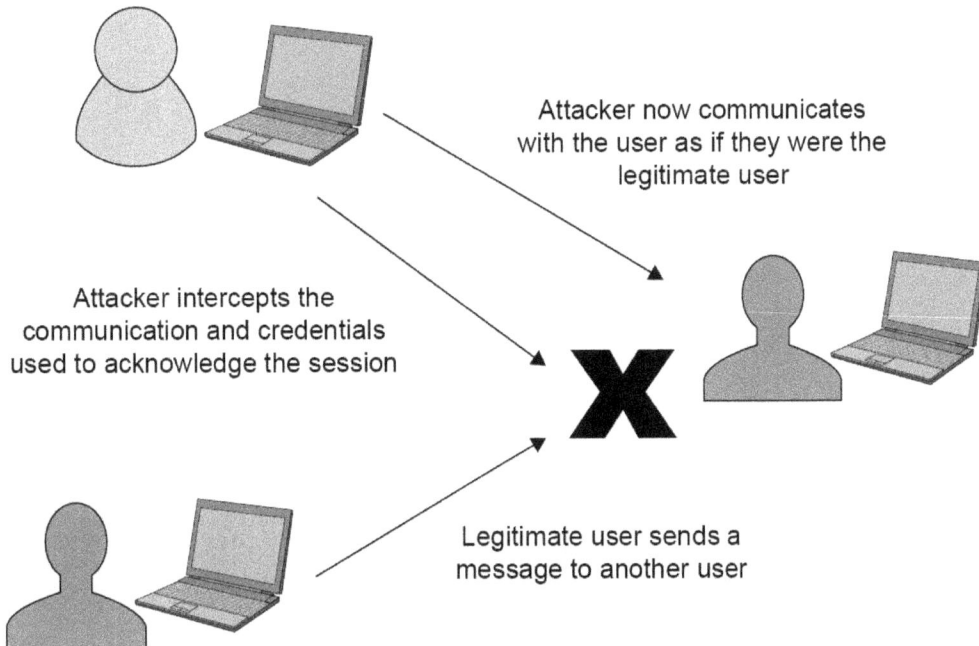

Figure 1.9: Attacker-in-the-middle (AiTM) attack: Attacker intercepts
and impersonates users in a communication

Proper user education on phishing email campaigns and having a zero-trust model for user authentication and access will help to protect against these types of attacks.

Injection Attacks – Application and Data Layers

Injection attacks are a threat primarily to databases that are connected to our applications. These threats are like brute-force attacks, as they make an active effort to gain access to systems. The way that injection attacks gain access is by sending a command or query to a database that takes advantage of a known flaw in the database. This command code or query is then executed without proper authorization, allowing the attacker to gain access to sensitive data.

Figure 1.10 illustrates the process of how this attack may take place on a SQL database:

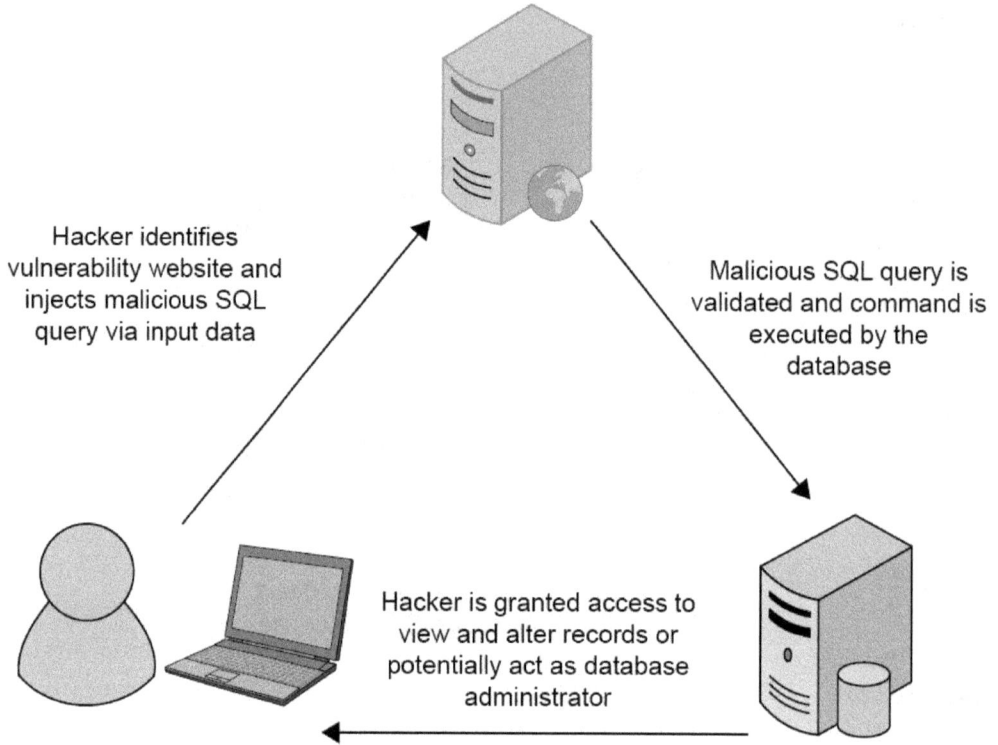

Hacker identifies vulnerability website and injects malicious SQL query via input data

Malicious SQL query is validated and command is executed by the database

Hacker is granted access to view and alter records or potentially act as database administrator

Figure 1.10: SQL injection attack: Hacker injects malicious SQL to access and manipulate database records

This injection attack is caused by poor authentication and monitoring controls for the database. *Figure 1.11* shows the process of how the attacker gains access to the user's session cookies:

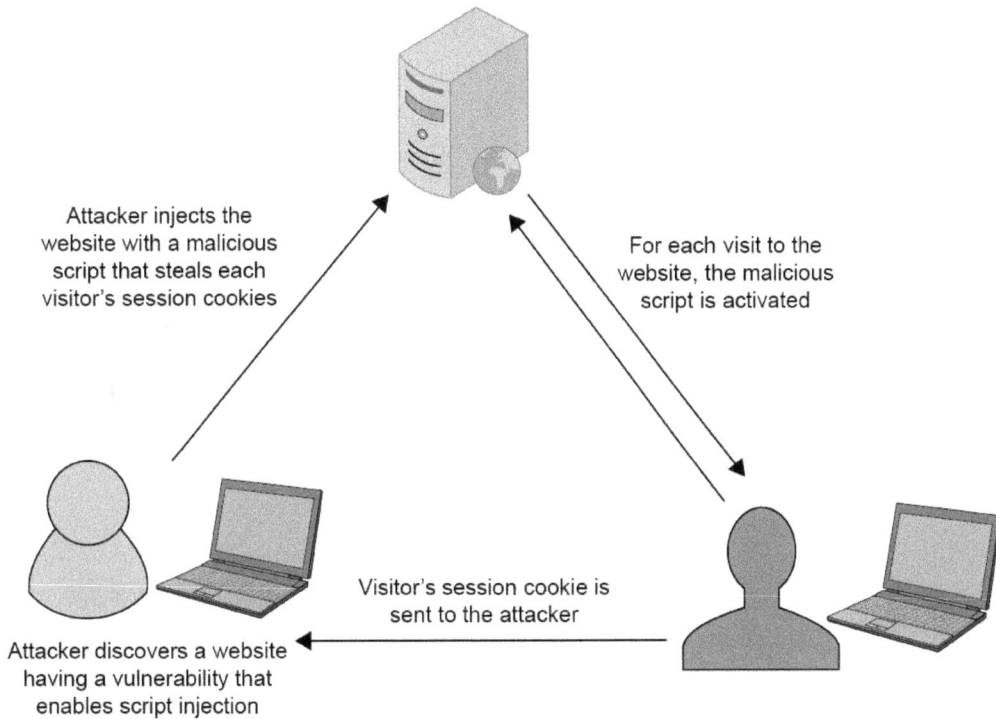

Attacker injects the website with a malicious script that steals each visitor's session cookies

For each visit to the website, the malicious script is activated

Visitor's session cookie is sent to the attacker

Attacker discovers a website having a vulnerability that enables script injection

Figure 1.11: Cross-site scripting (XSS) attack: An attacker injects a malicious script into a vulnerable website to steal visitors' session cookies

The visitor to the website has no knowledge that their session cookie has been intercepted and that they have been redirected. This allows the attacker to interact with the user's device and activate malicious code and malware.

> **Note**
> The external threats to companies and users are always evolving. A great resource to keep up with the most current risks is the OWASP Top 10 Web Application Security Risks: `https://owasp.org/Top10/`.

Likelihood			
Very likely	Acceptable risk medium 2	Unacceptable risk high 3	Unacceptable risk extreme 5
Likely	Acceptable risk low 1	Acceptable risk medium 2	Unacceptable risk high 3
Unlikely	Acceptable risk low 1	Acceptable risk low 1	Acceptable risk medium 2
What is the chance that it will happen?	Minor	Moderate	Major

Impact
How Serious is the Risk?

Figure 1.12: Security risk matrix visualized

In *Figure 1.12*, the x axis shows the impact scale, indicating the severity of the risk. The y axis shows the likelihood scale, representing the probability of the risk occurring. The intersection of these scores generates a risk score, guiding the prioritization of remediation efforts.

Throughout this book, we will discuss the ways that a cybersecurity architect can evaluate and design infrastructures to protect and remediate potential internal and external threats and vulnerabilities before they are exploited and turn into attacks.

Social Engineering

Social engineering attacks manipulate human behavior to gain unauthorized access to systems or information. Here are some common types and techniques:

- **Phishing**: Attackers send fraudulent emails or messages that appear to come from legitimate sources, tricking recipients into revealing sensitive information or clicking on malicious links.
- **Spear phishing**: A more targeted form of phishing, where attackers customize their messages to a specific individual or organization, often using personal information to appear more convincing.

- **Whaling**: Like spear phishing but targets high-profile individuals such as executives, aiming to gain access to sensitive corporate information.

- **Baiting**: Attackers offer something enticing, such as free software or a gift, to lure victims into providing personal information or downloading malware.

- **Pretexting**: Attackers create a fabricated scenario to trick victims into divulging information or performing actions that compromise security.

- **Quid pro quo**: Attackers promise a service or benefit in exchange for information or access, such as pretending to be IT support and offering help in return for login credentials.

- **Tailgating/piggybacking**: Attackers physically follow authorized personnel into restricted areas without proper authentication.

- **Vishing**: Voice phishing, where attackers use phone calls to trick victims into revealing personal information.

- **Smishing**: SMS phishing, where attackers send fraudulent text messages to deceive victims into providing sensitive information.

- **Honeytrap**: Attackers use romantic or seductive approaches to manipulate victims into sharing confidential information.

Understanding these techniques helps in developing effective security awareness programs and implementing measures to protect against social engineering attacks.

Defense in Depth: A Real-Life Example

So far in this chapter, we have discussed the various logical layers in a technology stack, the types of controls that you may choose to configure in some or all of these layers, and the types of attacks that your infrastructure might face, from internal and external sources.

What we have not yet done is brought this all together to demonstrate the real, tangible benefit of a defense-in-depth strategy.

Now that you know about the foundational components, it's time to refer to a real-world example of a successful cyber-attack by a persistent and well-resourced attacker, where you will gain an understanding of the stages of the attack and the opportunities for defense that could have prevented the attack from being successful or limited the impact from it.

On December 13, 2020, news broke in IT, security, and even mainstream news media of a breach of SolarWinds, a company that makes software to manage networks and IT infrastructure.

Cybercriminals believed to be working for a nation-state had managed to insert a backdoor into their widely used software, Orion. A backdoor is where a malicious actor inserts software code or a misconfiguration that allows them to gain access to the software of a system at will without being detected. It is called a backdoor because, though you might lock the front door of your home and have security measures, if you leave your back door open, you are vulnerable to intrusion.

The initial discovery was not made by SolarWinds, but by one of its customers, a computer security firm called FireEye. Their investigation led them to the conclusion that they had been breached using Orion, and that there was what looked like a backdoor in the Orion software.

An advisory issued by the U.S. Department of Homeland Security indicated that the affected versions of SolarWinds Orion were versions 2019.4 through 2020.2.1 HF1. In total, more than 18,000 SolarWinds customers installed malicious updates.

This attack is commonly referred to as a **supply chain attack** – the affected businesses were impacted by a compromise of software that they consume, from a vendor in their supply chain.

If you are wondering what the defense-in-depth angle is here, there were several ways in which customers could have avoided being vulnerable to the Orion compromise:

- Had they not exposed their Orion servers to the internet, the threat actors could not have compromised their systems by exploiting the backdoor – this is a network security layer control. **Microsoft Defender for Cloud** and **Microsoft Defender XDR** may have been able to detect these publicly accessible servers and alert them that they should be remediated.

- Had they been vigilantly updating and patching the software, they would not have been running the vulnerable versions for as long – this is an application layer control. **Microsoft Defender for Cloud** and **Microsoft Defender XDR** could have alerted about old software versions with known vulnerabilities. While this vulnerability was not yet known, all software has bugs and vulnerabilities – a regular update cadence helps ensure that known issues are resolved before they can be exploited.

Similarly, the SolarWinds compromise, which allowed the threat actors access to their systems to add the backdoor into their software without being detected, could have been either averted or detected sooner through the following:

- **Microsoft Defender for Cloud** may have been able to detect weaknesses in the software development processes in SolarWinds, such that the attempt to insert malicious code may have been detected.

- Had SolarWinds had more extensive logging, linked to a **security information and event management** (SIEM) platform such as **Microsoft Sentinel**, they may have been able to spot unusual or malicious behavior in their systems.

- The initial access vector for compromising SolarWinds was a compromised **virtual private network** (**VPN**) account belonging to one of their employees. Had SolarWinds been monitoring credentials, the behavior of identities, and other signals, they might have identified earlier that a malicious attacker had gained unauthorized access to their network – this is the identity and access layer. Products such as **Microsoft Defender for Identity** and **Microsoft Entra Conditional Access** are designed to surface insights such as this.

What you should take away from this chapter and this very high-level summary of a now infamous attack is that no single security control is infallible; you need as many security controls in as many layers of your infrastructure as you can afford and are reasonable given the risk appetite of your business.

If a single control fails, other controls may yet prevent a successful compromise, reduce lateral movement, or alert you sooner to unauthorized activity.

> **Note**
>
> You can read the full, detailed writeup on Wired.com at this link: `https://www.wired.com/story/the-untold-story-of-solarwinds-the-boldest-supply-chain-hack-ever/`.

Additional Example: Okta

In October 2023, Okta experienced a significant security incident that underscores the importance of defense in depth. What follows is a detailed summary of the incident, its impact, response, and remediation, highlighting where defense in depth could have mitigated the breach.

Initial Signs of Compromise

The breach began when attackers used stolen credentials to access Okta's customer support system. The initial signs of compromise were detected when unusual activity was observed in the support system, specifically unauthorized access to **HTTP Archive** (**HAR**) files uploaded by customers.

Impact

The attackers gained access to sensitive data, including authentication tokens and personal information of Okta's customer support users. This breach affected nearly all customer support users, exposing names, email addresses, and other sensitive details. Approximately 1% of Okta's 18,000+ customers had their authentication tokens stolen, which could be used to alter customer accounts.

Impact

Okta's response involved several critical steps:

1. **Engagement with law enforcement**: Okta promptly notified law enforcement agencies to investigate the breach.

2. **Customer notifications**: Affected customers were informed, and Okta provided a customized impact report along with recommendations to mitigate potential phishing and social engineering attacks.

3. **Publication of indicators of compromise (IOCs)**: Okta shared IOCs to help customers identify and respond to similar threats.

4. **Enhanced security measures**: Okta reviewed and enhanced the security of its support system, including changes to access provisioning and data retention policies.

Remediation

To prevent future incidents, Okta implemented several remediation measures:

1. **Zero standing privileges**: Admin roles are now requested, approved, and assigned only for the duration needed.

2. **Multi-factor authentication (MFA)**: MFA is required for critical actions in the admin console.

3. **Dynamic zones**: Okta introduced the ability to detect and block requests from anonymizers to protect critical assets.

4. **IP binding**: Sessions are invalidated if the source IP changes during the session, preventing session takeover.

5. **Allowlisted network zones for APIs:** This restricts attackers from stealing and replaying tokens outside specified IP ranges.

Defense in Depth

The breach could have been mitigated or even avoided with stronger defense-in-depth strategies:

1. **Credential management**: Implementing stricter controls on credential storage and usage could have prevented the initial compromise – for example, ensuring that service account credentials are never stored on personal devices.

2. **MFA**: Enforcing MFA for all administrative accounts would have added an additional layer of security, making it harder for attackers to gain access even with stolen credentials.

3. **Network segmentation**: Isolating the support system from other critical systems could have limited the attackers' ability to move laterally and access sensitive data.

4. **Regular audits and monitoring**: Continuous monitoring and regular security audits could have detected the unusual activity sooner, allowing for a quicker response.

This incident highlights the critical need for a multi-layered security approach to protect against sophisticated cyber threats. By implementing defense-in-depth techniques, organizations can significantly reduce the risk and impact of security breaches.

> **Note**
>
> You can read more about this incident and response at `https://sec.okta.com/harfiles`.

Summary

In this chapter, we discussed multiple areas to consider as a cybersecurity architect within cloud and hybrid infrastructures. This included the variations in cybersecurity for on-premises data centers versus moving to cloud environments. As you move on to the next sections of this book and begin to determine how Microsoft's capabilities can be used to design a cybersecurity architecture for a company, these concepts and topics will be important to reference.

The key takeaways from this chapter are that there are a wide variety of TTPs employed by attackers to disrupt access to or gain unauthorized access to systems and data, both on-premises and in the cloud. These attacks can occur across multiple layers of your technology stack, and attackers often chain TTPs together, exploiting small vulnerabilities to create significant risks. Also, a defense-in-depth approach is essential for securing your systems and data. Finally, Microsoft offers several products designed to detect and protect against these risks across various layers.

The next chapter will discuss how to build an overall security strategy and architecture with a focus on the **Microsoft Cybersecurity Reference Architectures**.

Exam Readiness Drill – Chapter Review Section

Apart from mastering key concepts, strong test-taking skills under time pressure are essential for acing your certification exam. That's why developing these abilities early in your learning journey is critical.

Exam readiness drills, using the free online practice resources provided with this book, help you progressively improve your time management and test-taking skills while reinforcing the key concepts you've learned.

How to Get Started

1. Open the link or scan the QR code at the bottom of this page.
2. If you have unlocked the practice resources already, log in to your registered account. If you haven't, follow the instructions in *Chapter 11* and come back to this page.
3. Once you have logged in, click the **START** button to start a quiz.

We recommend attempting a quiz multiple times till you're able to answer most of the questions correctly and well within the time limit.

You can use the following practice template to help you plan your attempts:

Working On Accuracy		
Attempt	Target	Time Limit
Attempt 1	40% or more	Till the timer runs out
Attempt 2	60% or more	Till the timer runs out
Attempt 3	75% or more	Till the timer runs out
Working On Timing		
Attempt 4	75% or more	1 minute before time limit
Attempt 5	75% or more	2 minutes before time limit
Attempt 6	75% or more	3 minutes before time limit

The above drill is just an example. Design your drills based on your own goals and make the most of the online quizzes accompanying this book.

First time accessing the online resources? 🔒
You'll need to unlock them through a one-time process. **Head to** *Chapter 11* **for instructions**.

Open Quiz

https://packt.link/SC100_CH01

Or scan this QR code →

Build an Overall Security Strategy and Architecture

This chapter will discuss the ways that a **cybersecurity architect** delivers the development and design of a security strategy. Cybersecurity architects must also be able to communicate that strategy to business leaders and provide direction on how it integrates cloud, hybrid, and multi-tenant environments. Once you have completed this chapter, you will have the tools and knowledge to identify, design, and communicate these requirements to the company.

This chapter extends on the **Design solutions that align with security best practices and priorities** domain, which constitutes **20–25%** of the SC-100 Exam Guide. You'll learn how to develop a resiliency strategy for ransomware and other attacks based on Microsoft security best practices. Additionally, this chapter covers designing solutions that adhere to the **Microsoft Cybersecurity Reference Architectures (MCRA)** and the **Microsoft Cloud Security Benchmark (MCSB)**. Furthermore, it provides insights into designing solutions that align with the Microsoft Cloud Adoption Framework for Azure and the Microsoft Azure Well-Architected Framework.

In this chapter, you are going to learn about the following main topics:

- Identifying the integration points in an architecture by using the Microsoft Cybersecurity Reference Architectures

- Translating business goals into security requirements

- Translating security requirements into technical capabilities

- Designing security for a resiliency strategy

- Integrating a hybrid or multi-tenant environment into a security strategy

- Developing a technical and governance strategy for traffic filtering and segmentation

Identifying the Integration Points in an Architecture by Using the Microsoft Cybersecurity Reference Architectures

Cybersecurity architects must have a solid foundation in the design and integration of security across multiple environments. As is the case with any design, planning is necessary at the beginning before any formal architecture can be created and performed. The cybersecurity architect must understand the current environment and how the various components of that environment interact and integrate. This includes the controls that are currently in place and how they are used.

Cybersecurity architects will be involved in designing security architectures based on best practices and security frameworks, as well as existing company environments. You may be asked to retrofit security controls from an on-premises environment to cloud environments that are part of a company's journey to the cloud. This requires a broad understanding of not just Microsoft technology but third-party and other cloud provider security controls and capabilities.

Microsoft has assisted in this process by creating the **MCRA**. The MCRA provides diagrams and templates that can assist the cybersecurity architect with identifying and planning for the various integration points of on-premises and multi-cloud security controls into Microsoft security posture management and security operations solutions, such as **Microsoft Defender for Cloud**, **Microsoft 365 Defender**, and **Microsoft Sentinel**. The MCRA, like other reference architecture diagrams and guides from Microsoft, are exactly what they are stated to be, for reference. They are not expected to be a comprehensive guide that covers every situation and scenario that a cybersecurity architect will uncover in their design, but they are a good starting point.

> **Note**
> The MCRA and the various diagrams and templates can be found at this link: `https://learn.microsoft.com/en-us/security/adoption/mcra`.

The following sections will provide additional information on the components of the MRCA and how a cybersecurity architect can use these tools to determine the various integration points and controls within the security design.

How is the MCRA Used?

The MCRA provides guidance for a cybersecurity architect as well as others within the company, including security operations, compliance administrators, and security administrators, with overviews and materials to help with the overall security, compliance, and identity strategy of the company. The MCRA can be used in diverse ways:

- Providing a starting point or template for the initial design
- Providing a comparison of the current security architecture for evaluation and recommendations

- Learning about cybersecurity through the references and content within the MCRA

- Learning about Microsoft's capabilities for security, compliance, and identity

- Learning about the investments required for integrating Microsoft solutions into the security architecture

Now that you know some ways in which the MCRA can be used, let us move on to the next section, which will further explain the components of the MCRA.

What Are the Components of the MCRA?

The MCRA is a set of diagrams, templates, and links that assist in the understanding and planning of a cybersecurity architecture that is integrated primarily with Microsoft. Even though the architecture diagrams focus on Microsoft solutions, they are helpful to the overall understanding of a complete cybersecurity architecture.

Figure 2.1 provides an overview of the MCRA and the sections of the architecture:

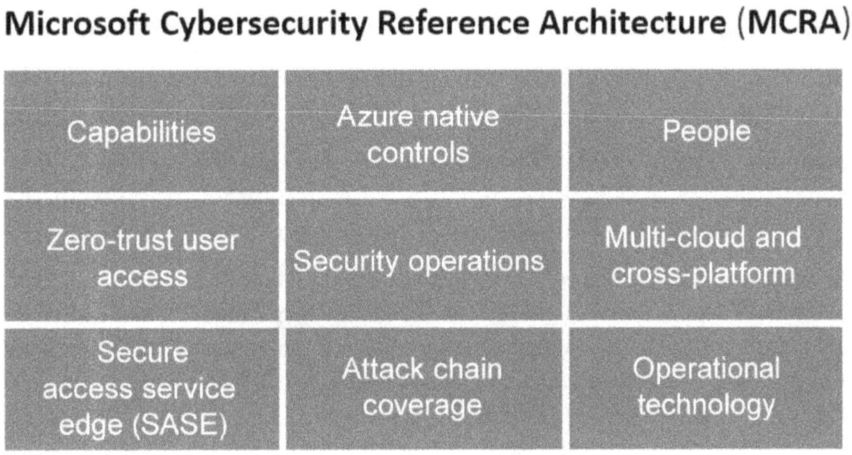

Microsoft Cybersecurity Reference Architecture (MCRA)

Capabilities	Azure native controls	People
Zero-trust user access	Security operations	Multi-cloud and cross-platform
Secure access service edge (SASE)	Attack chain coverage	Operational technology

Figure 2.1: MCRA topics

The areas that the MCRA focuses on are resources, capabilities, controls, access, and operations:

- **Capabilities**: These are the capabilities that Microsoft has for security, compliance, and identity architecture for monitoring, managing, and operating a secure environment for public, hybrid, and multi-cloud infrastructures. As a cybersecurity architect, you should be familiar with these capabilities to provide a solid architecture for security, compliance, and identity.

- **People**: This is the most important asset of a company. As a cybersecurity architect, you should design an architecture that protects people from threats through a clear definition of roles and responsibilities. This could be identity theft, personal information, and reputation through the exposure of resources that are within the company applications and data. Reputational damage also plays a role within the overall company and its board members if there is a cybersecurity event.

- **Zero-trust user access**: Zero trust is the concept of requiring constant verification of resources for access. This includes user access. Enforcing zero trust is a way to protect the people who are accessing resources from cybersecurity threats and stolen identities.

- **Attack chain coverage**: *Chapter 1, Cybersecurity in the Cloud*, discussed the cybersecurity attack chain and how designing a defense-in-depth architecture addresses these areas of the attack chain. The capabilities of **Microsoft Defender** solutions, **Microsoft Insider Risk Management**, **Microsoft Defender External Attack Surface Management** (**Microsoft Defender EASM**), and **Microsoft Sentinel** help to address this attack chain within Microsoft, hybrid, and multi-cloud infrastructures.

- **Security operations**: **Security information and event management** (**SIEM**) and **security orchestration, automation, and response** (**SOAR**) solutions such as Microsoft Sentinel build a solid foundation for a company to build a **security operations center** (**SOC**). The strength of a SOC is built on the ability to recognize and respond to threats and incidents. The cybersecurity architect will assist in the evaluation and use of the tools available.

- **Operational technology** (**OT**): This addresses the many connected devices within a company and how to design an architecture for these devices that utilizes zero trust. This helps these devices stop being gateways for attackers to gain access to critical applications and data.

- **Azure native controls**: Microsoft provides many native controls within their solutions for Microsoft 365 and Azure. This includes Microsoft Entra capabilities for zero-trust enforcement, such as Conditional Access policies and Entra ID Protection. These Azure native controls are also found in many of the **platform-as-a-service** (**PaaS**) offerings for application services and databases with automated security patching and backups. Storage and databases are encrypted at rest by default to protect against threats. Microsoft Defender for Cloud is also enabled by default with auditing policies to provide guidance for building a strong security posture.

- **Multi-cloud and cross-platform**: Microsoft's security, compliance, and identity protection tools are not just for Azure services. These capabilities can be utilized across Microsoft 365, Dynamics 365, Power Platform, SharePoint Online, and the entire Microsoft cloud solutions. In addition, Microsoft has built its solutions to be utilized for hybrid architectures with on-premises resources as well as multi-cloud infrastructures that utilize **Amazon Web Services** (**AWS**) and **Google Cloud Platform** (**GCP**). These also include the use of **Azure Arc** to enable simpler and faster integration with multi-cloud infrastructure. A cybersecurity architect can design an architecture in Microsoft that can address the entire security, compliance, and identity needs of a company.

- **Secure access service edge** (**SASE**): The capabilities of Microsoft for Azure native controls, support for multi-cloud and cross-platform, and IoT OT allow for the creation of a SASE architecture. SASE utilizes these solutions and zero-trust methodologies to harden endpoints across the company architecture and monitor for threats and vulnerabilities.

Zero trust is a very key component in the overall cybersecurity architecture. Microsoft has developed a zero-trust **rapid modernization plan** (**RaMP**) within the MCRA that can be used as the foundation for designing a zero-trust architecture. Zero trust as a concept means that a breach should always be assumed and that you should constantly verify access to resources. You need to address zero trust throughout your cybersecurity architecture in how you protect identities, devices, networks, applications, data, and compute infrastructure. Each of these areas will be addressed throughout the design of a cybersecurity architecture.

All the components of the MCRA help to enforce a zero-trust architecture and address the Cyber Kill Chain by hardening company resources and monitoring for threats and vulnerabilities. A Microsoft cybersecurity architect should be extremely familiar with the MCRA to provide them with a foundation and guidance for designing security, compliance, and identity within Microsoft 365, Azure, hybrid, and multi-cloud infrastructures.

The next section will discuss how a cybersecurity architect will work with the business to address security requirements that align with the company's business goals.

Translating Business Goals into Security Requirements

Cybersecurity, as is the case with most **information technology** (**IT**), is thought of by businesses as unavoidable and something that must be implemented. Therefore, a cybersecurity architect needs to be able to align the controls to protect the company with the business goals of the company.

Cyber-attacks affect companies in diverse ways. They can damage the reputation of the victim company in the case of a high-profile attack, cause economic damage if important business documents or financial data are breached, and lead to regulatory costs from potential fines if the breach is caused by inadequately addressing compliance standards.

Each of these business impacts needs to be addressed and presented to the company when building a cybersecurity architecture. A proper risk analysis should be done for threats to identify proper security controls and present them to the company to help them understand the security and business impact.

Threat Analysis

Threat analysis is a process that every company should go through to properly understand the vulnerabilities within their systems and the potential threat to the company if those vulnerabilities are exploited.

When doing a threat analysis, you need to consider the perceived risk to an asset. To accomplish this, you must consider the combination of the asset along with any vulnerabilities and the potential threat that the vulnerability will be exploited. As an equation, this would look like this:

Asset (A) + Vulnerability (V) + Threat (T) = Risk (R)

The assessment of the risks to assets can be categorized into various levels of risk, as shown in *Figure 2.2*:

Likelihood		Minor	Moderate	Major
	Very likely	Acceptable risk medium 2	Unacceptable risk high 3	Unacceptable risk extreme 5
	Likely	Acceptable risk low 1	Acceptable risk medium 2	Unacceptable risk high 3
	Unlikely	Acceptable risk low 1	Acceptable risk low 1	Acceptable risk medium 2
	What is the chance that it will happen?	Minor	Moderate	Major

Impact
How Serious is the Risk?

Figure 2.2: Risk assessment matrix

Identifying the company assets and then assessing the risk should be done at the business and technical level so that you have a proper understanding and perception across departments.

Once all the assets and levels of risk have been identified, then the analysis of financial exposure needs to be determined. This can be accomplished by determining the following criteria for each asset:

- **Exposure factor** (**EF**) is the impact that is measured by the percentage loss of an asset if the risk is realized.

- **Single loss expectancy** (**SLE**) is the value of the asset multiplied by the EF. This will place a financial value on the asset loss when exposed.

- The **annualized rate of occurrence** (**ARO**) is the possible number of times that this risk may be exploited over the year.

- The **annualized loss expectancy** (**ALE**) is the combined financial impact of the SLE times the ARO, thus providing an annual cost of loss for the asset.

Calculating the ALE for all the company's digital assets places a tangible value on those assets and then allows a company to evaluate the cost of investing in the controls needed to mitigate or avoid that risk. If a company is not going through these steps, they are not taking due care and due diligence to protect company assets.

Another point to make here is that this is placing a financial value on assets and not reputational value. The reputational impact of data exposure or data loss can be much more damaging to a company with its customers than the perceived value of the assets. Companies should consider this when determining investment in security controls as well.

> **Note**
>
> The following is a valuable resource if you would like to learn more about cyber threat analysis: https://csrc.nist.gov/pubs/sp/800/30/r1/final.

The cybersecurity architect should work with the business in agreeing to the priority of controls to implement and budgeting the roadmap to a complete cybersecurity design. A company that has previously performed a NIST or Microsoft Security Adoption Framework analysis should look at the Microsoft capabilities and the total cost of ownership for licensing that they may already have with Microsoft 365 and Azure.

The importance of aligning business goals to security requirements should not be understated. As is true with any cloud adoption, the successful implementation of security controls will be measured by how they address the business goals and needs of the company. The cybersecurity architect should always be thinking about these key indicators of success when designing a cybersecurity architecture.

The next section will discuss Microsoft's security products and services and how to utilize them to address a company's security requirements.

Translating Security Requirements into Technical Capabilities

Now that you understand the security posture, defense-in-depth, and shared responsibility as you begin to architect cybersecurity for the cloud, it is time to discuss the makeup of a security operations team and the levels of a cybersecurity attack.

In *Chapter 1*, *Cybersecurity in the Cloud*, the concept of building a defense-in-depth security strategy was discussed. In this section, you will take each of those defense-in-depth requirements and align them with some security products, services, and processes. *Figure 2.3* shows the defense-in-depth strategy:

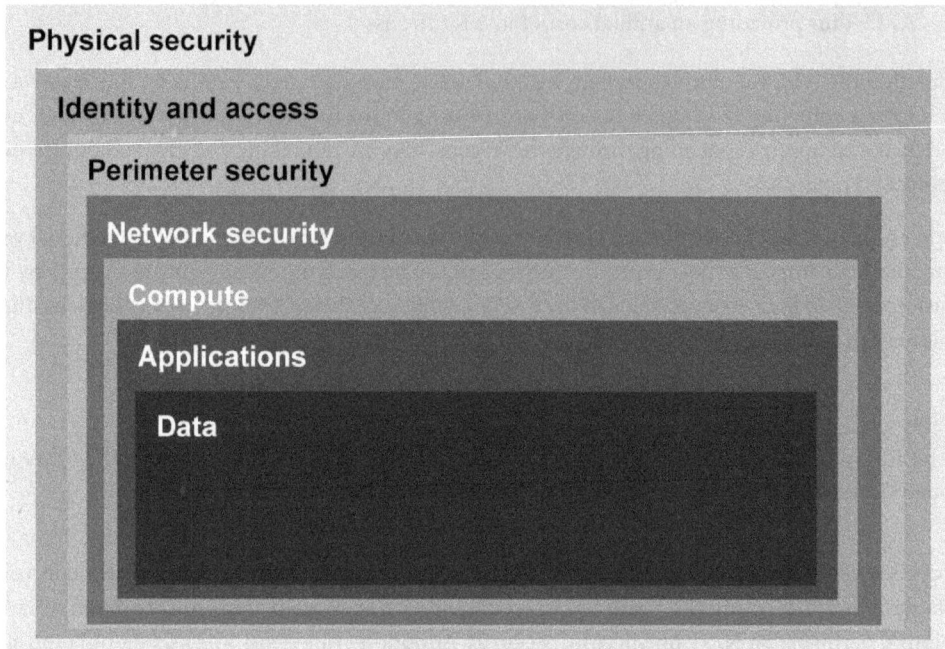

Figure 2.3: Defense-in-depth security diagram

Each of these layers can be protected with controls that a cybersecurity architect should address in the design.

Physical

The **physical** level of defense includes the actual hardware technology and spans the entire data center facility. In a cloud infrastructure, the cloud provider, such as Microsoft, protects this layer. However, if your company has a hybrid infrastructure with on-premises resources, then the security of this layer cannot be forgotten.

> **Note**
>
> More information on physical security within Microsoft data centers can be found at this link: https://docs.microsoft.com/en-us/azure/security/fundamentals/physical-security#physical-security.

Hybrid Security within a Physical Data Center

In a hybrid infrastructure model, Microsoft can provide additional controls, such as **mobile device enrollment** (MDE) with USB management controls. It can also provide controls for the on-premises infrastructure with **Microsoft Defender for Endpoint** and **Microsoft Defender for Identity**. Also, it can use **Microsoft Defender for Cloud** with **Azure Arc** for on-premises virtual machines and containers.

Identity and A   ccess

The more cloud services that you begin to utilize as a company, the more important identity becomes as the first line of defense for a cybersecurity architect to address. Microsoft has introduced **Microsoft Entra** as the foundation for identity and access management and protection. Microsoft Entra includes Entra ID for identity and access and Permissions Management for governance and verified ID.

> **Note**
> The following link provides more information on Microsoft Entra: `https://docs.microsoft.com/en-us/entra/`.

Identity and access protection, such as **Conditional Access policies**, provides a foundation to protect against identity theft and vulnerabilities through making verification decisions based on the conditions of users, devices, and locations:

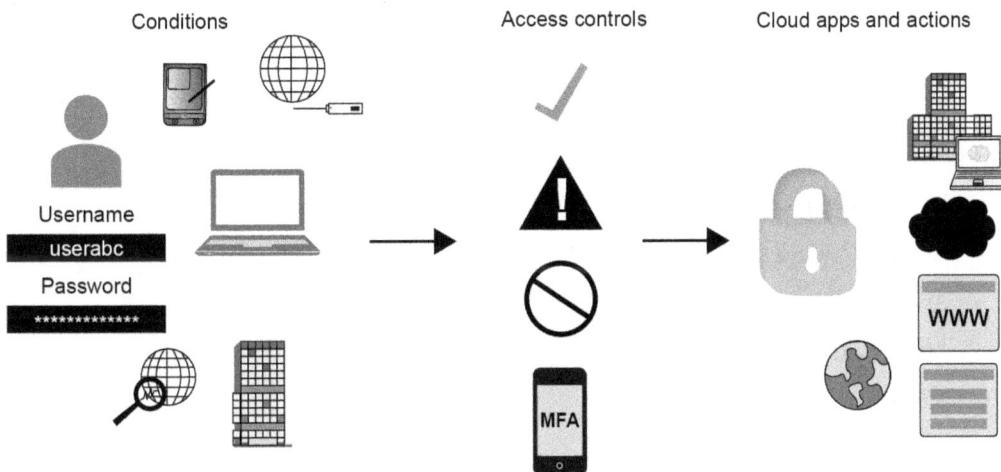

Figure 2.4: Zero trust with Conditional Access

Figure 2.4 shows how Conditional Access policies can be used to enforce additional verification for access.

Perimeter security

Within Microsoft, the **Azure DDoS Protection** service provides a higher level of reporting and customization for a company than the basic free service. *Figure 2.5* shows how Azure DDoS Protection is enforced for perimeter security:

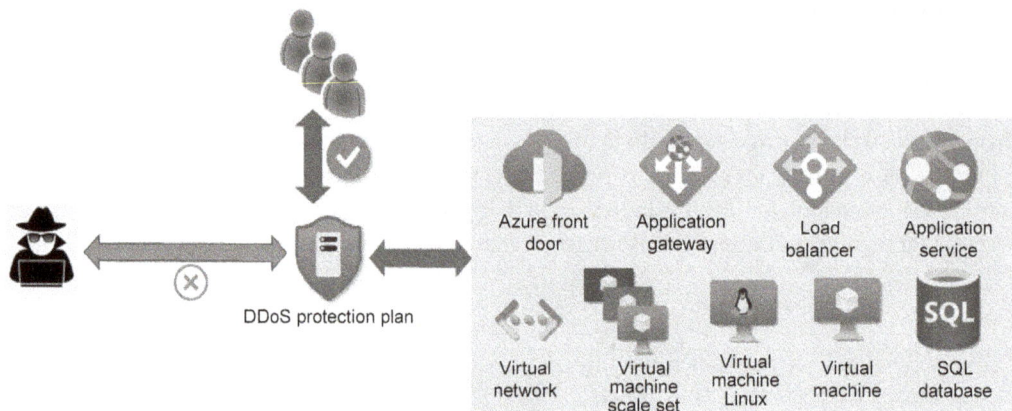

Figure 2.5: Azure DDoS Protection

There are two tiers of service offered within Azure DDoS Protection.

DDoS Network Protection

Network protection is billed per 100 protected IP addresses and provides the following features:

- Active traffic monitoring and always-on detection
- L3/L4 automatic attack mitigation
- Automatic attack mitigation
- Application-based mitigation policies
- Metrics and alerts
- Mitigation reports
- Mitigation flow logs
- Mitigation policies tuned to customer's application
- Integration with Firewall Manager
- Microsoft Sentinel data connector and workbook

- Protection of resources across subscriptions in a tenant

- Public IP standard tier protection

DDoS IP Protection

In addition to the features of DDoS network protection, the DDoS IP protection tier offers the following features and is billed on a per IP address basis:

- Public IP basic tier protection

- DDoS rapid response support

- Cost protection

- WAF discount

For additional perimeter protection, the company can implement network security groups, virtual firewall appliances, or **Azure Firewall** to protect the tenant's perimeter to block port and packet-level attacks. Premium Azure Firewall services provide intrusion detection and intrusion protection capabilities, as well as an **application gateway** with a **web application firewall** (WAF) to protect from application-layer attacks. In addition, **Azure Virtual WAN** can provide a level of protection to isolate virtual connections within Azure and on-premises from potential threats through the internet:

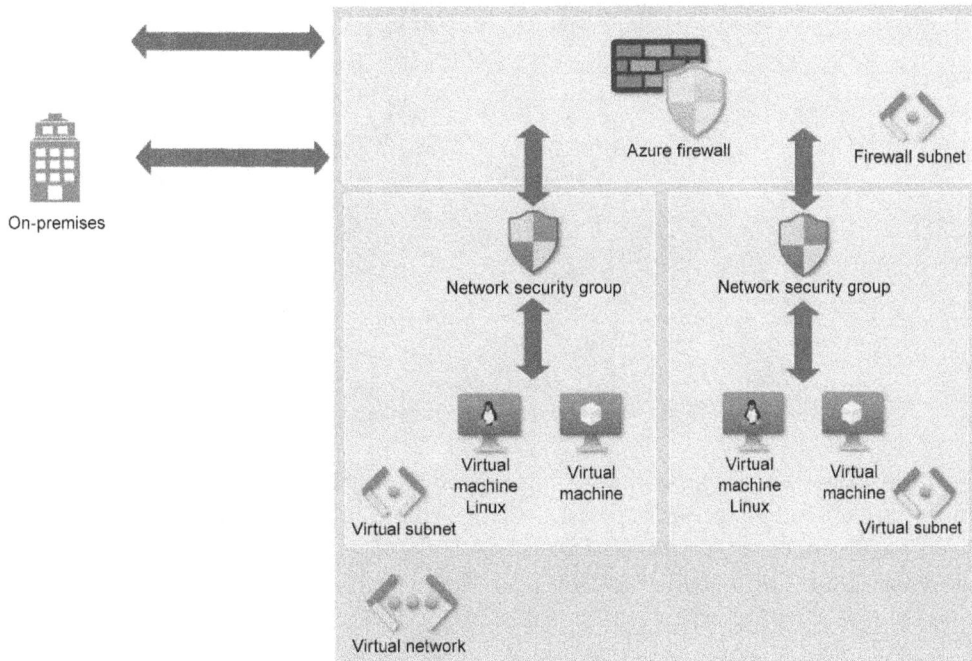

Figure 2.6: Azure Firewall

Figure 2.6 shows how Azure Firewall can direct traffic at the perimeter for internet and on-premises traffic.

Network Security

In an Azure infrastructure, **network security groups** and **application security groups** can also be configured on network interfaces with additional ports, IP addresses, or application-layer rules for how traffic can be routed within the network. Service endpoints can be configured within network and application security groups to make sure that traffic is not routed outside of the Azure infrastructure:

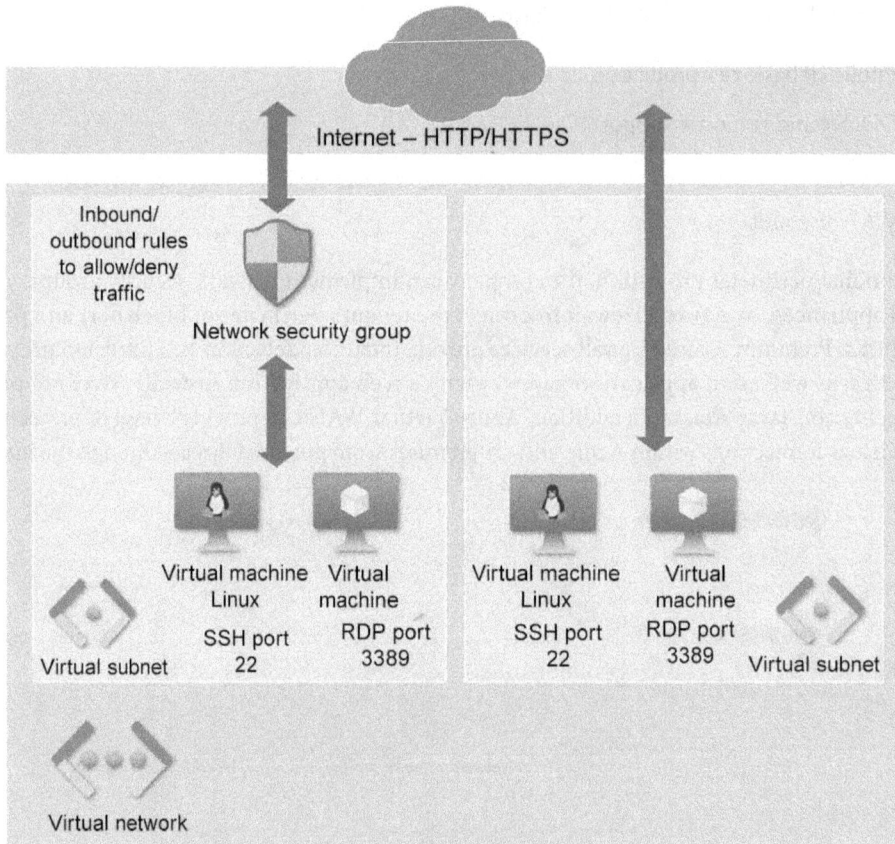

Figure 2.7: Network security groups for network perimeter protection

Figure 2.7 shows how a subnet network security group can allow and deny network perimeter traffic. It depicts north/south traffic between virtual machines within one of two subnets in a virtual network, with a network security group in place on one of the subnets to control inbound and outbound traffic. Some standard management ports are shown – SSH port 22 for Linux and RDP port 3389 for Windows.

Compute

A common attack at the compute layer involves scanning and gaining access to management ports on devices. Not exposing these ports, via 3389 for the Windows **Remote Desktop Protocol (RDP)** and 22 for the Linux **Secure Shell (SSH)** protocol, to the internet will provide a layer of protection against these attacks. Within Microsoft Azure, this can be accomplished with network security group rules, removing public IP addresses on virtual machines, **bastion hosts**, and/or utilizing **just-in-time (JIT)** virtual machine access. Many of these security options will be discussed later in this book:

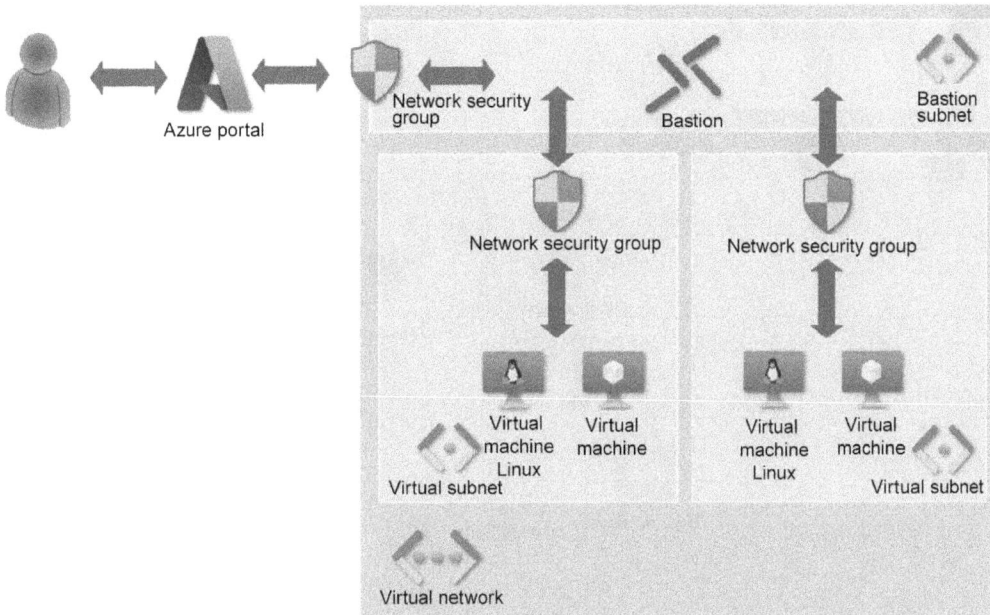

Figure 2.8: Azure Bastion for virtual machine remote management

Figure 2.8 shows how Azure Bastion isolates traffic on remote management ports to protect access to virtual machines.

Applications

Application workloads can be protected by utilizing Microsoft Defender for Containers and Microsoft Defender for App Service. In addition, Microsoft Defender for Cloud Apps assists in discovering applications that may introduce vulnerabilities and threats to your company. *Figure 2.9* shows how a WAF works to protect application-layer traffic from common application threats as defined in the OWASP Top 10.

> **Note**
>
> More information on the OWASP Top 10 can be found at `https://owasp.org/www-project-top-ten/`.

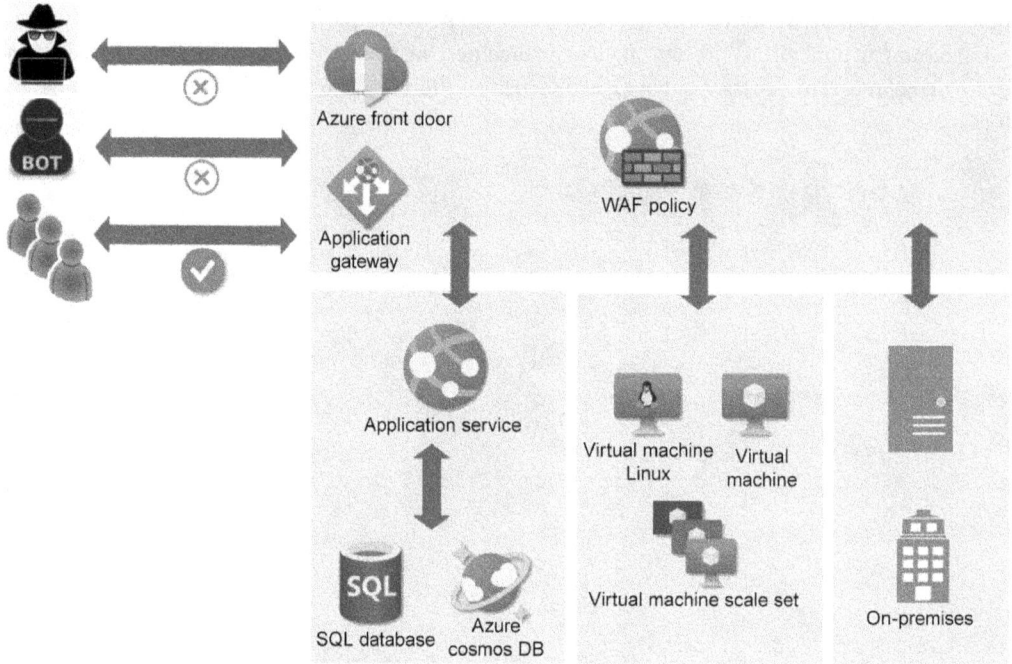

Figure 2.9: WAF diagram

The OWASP Top 10 provides a list of the most prominent security risk categories, including broken access control, security misconfiguration, and cryptographic failures within applications. Azure WAF uses a set of managed WAF rules sets from the OWASP **Core Rule Set** (**CRS**) to provide protection against attacks from the OWASP Top 10.

> **Note**
>
> More information can be found at `https://coreruleset.org/`.

Data

Microsoft provides encryption at rest by default on all storage accounts through **Storage Service Encryption (SSE)**. Databases use **transparent data encryption (TDE)** for encryption at rest. You should utilize **Transport Layer Security/Secure Socket Layer (TLS/SSL)** encryption for data that is in transit to avoid data from being intercepted and read during transmission. Encryption involves converting data into ciphertext to prevent unauthorized access:

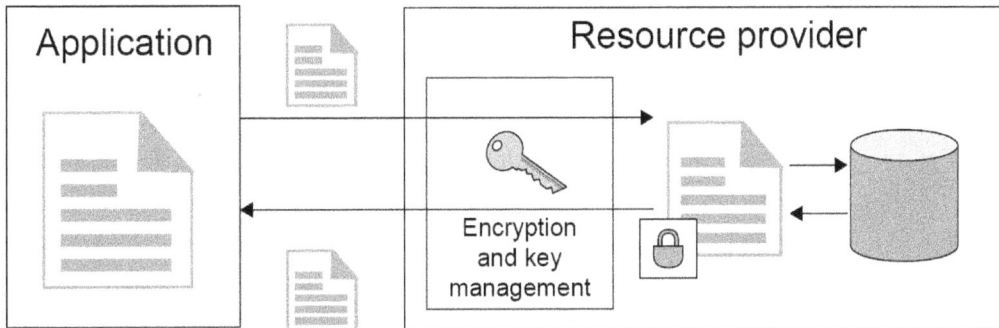

Figure 2.10: Data encrypted at rest

Figure 2.10 shows the process of encrypting data at rest. Only authorized users can view the data after they have authenticated.

In the next section, you will cover how to design security within a resiliency strategy.

Designing Security for a Resiliency Strategy

Not everything about cybersecurity involves protecting against threats and vulnerabilities. A key part of a cybersecurity architecture is having resiliency within the design. A resilient architecture will protect a company against disruption or data loss from an attack. Building a resilient architecture is within the network, compute, and storage design. Creating a resilient strategy for your architecture provides a level of business continuity to your company. This can allow continued operations in the case of a disastrous event.

> Note
>
> NIST 800-160 provides some guidance on resiliency and business continuity here: `https://csrc.nist.gov/pubs/sp/800/160/v2/r1/final`.

Resilient networks are built within virtual networking and security designed to provide both segmentation and redundancy throughout the architecture. Maintaining segmentation between private and public resources while also creating geographic separation and redundancy will provide resiliency in the case of an attack. When an attacker gains access to a network, security operations need to be able to create a barrier between the network that the threat actor or malicious intruder has accessed and the networks used for sensitive information that could be damaging to the company if the attacker gains access. A cybersecurity architect should take this into account when developing a design for the overall network architecture.

In addition to building network resiliency, the cybersecurity architect should also plan for compute resiliency. This resiliency utilizes both geographic redundancy and a strong strategy for backing up infrastructure resources. In Microsoft Azure, Azure Backup and Azure Site Recovery provide the backup and resiliency for Azure, as well as a hybrid infrastructure. Database services, such as Azure SQL Database and Cosmos DB, have these backup and resiliency capabilities built into them.

Storage resiliency is another area that a cybersecurity architect needs to consider to prevent data loss from an attack. Azure storage accounts provide geographic resiliency through geo-replication strategies. This alone does not provide resiliency; you should also have a strategy for backing up storage accounts and utilizing services such as Azure Key Vault to separate encryption keys, secrets, and certificates from Microsoft-managed keys within the storage account. Enabling soft-delete functionality prevents the accidental or intentional deletion of files and folders. Microsoft 365 provides a data loss prevention solution that can be used with Microsoft Purview on Azure Storage, Azure SQL Database, and AWS S3 storage accounts for additional resiliency protection to avoid the oversharing of company information.

Building a resilient architecture is one of the primary defenses against an attack such as ransomware. The resilient and redundant architecture provides recoverability to the company without losing access to critical applications and data.

A cybersecurity architect needs to understand the areas of vulnerability within the network and infrastructure and address them by recommending and designing resiliency. The following section will show you how to integrate hybrid and multi-tenant environments into the security architecture.

Integrating a Hybrid or Multi-Tenant Environment into a Security Strategy

This chapter has discussed several strategies and ways to implement zero trust, defense-in-depth, and resiliency in the architecture. At the time of writing, most companies have more than just Microsoft services and solutions within their infrastructure. Therefore, a cybersecurity architect needs to account for those additional environments in their design for the company. This includes private or on-premises data centers, other cloud providers (AWS or GCP), and multi-tenant Microsoft environments. Let us look at each of these and how to integrate them into the security strategy:

- **Private data centers**: Most companies today have a combination of private and public infrastructures also known as hybrid infrastructures. These architectures create complexity in network communications, identity and access management, and applications. The challenge is that these legacy private data centers have existing security controls in place. These controls were investments that the company previously made for physical and virtual protection. A cybersecurity architect needs to create a set of security controls based on organizational requirements and align them with Microsoft controls. Many of these controls will protect the physical layers of the data center, but other controls, such as firewalls, network virtual LANs, intrusion detection and prevention systems, and local data storage and encryption, also need to be integrated or replaced with Microsoft controls. The cybersecurity architect needs to balance these investments and their continued use within a hybrid architecture.

- **Other cloud providers**: AWS and Google have controls within their platforms for security monitoring and protection. Microsoft has created connectors and capabilities to use these tools and connect them to Microsoft Defender for Cloud, Azure Arc, and Microsoft Sentinel. This will be discussed in further detail in *Chapter 6, Evaluate the Security Posture and Recommending Technical Strategies to Manage Risk*, and *Chapter 3, Design a Security Operations Strategy*, respectively. The cybersecurity architect must take inventory of these controls and capabilities for the other cloud providers and determine the tools that can be used that are native to each platform and the integration points to Microsoft solutions.

- **Multi-tenant Microsoft environments**: Many enterprises have hybrid infrastructures as well as multiple Microsoft tenants. This could be from mergers and acquisitions, globally operated companies, or product segmentation. Each can have security controls and requirements based on local regulations and standards or product and service requirements. Again, the cybersecurity architect's role is to determine the controls that are in place, how they are being used by the company, and whether they should be used across all tenants or only with the current tenants. Some controls may only be used for a certain tenant. To prevent duplication and additional costs, the design should utilize and leverage solutions with multi-tenant capabilities, when available.

Now that you have covered some of the integration points, it is time to discuss traffic filtering and segmentation within the architecture in the next section.

Developing a Technical and Governance Strategy for Traffic Filtering and Segmentation

The last section of this chapter addresses how you would develop a strategy for traffic filtering and segmentation. In the *Translating Security Requirements into Technical Capabilities* section, you learned about protecting the network perimeter through network segmentation by utilizing different virtual networks and filtering traffic with Azure Firewall or a WAF.

As a cybersecurity architect, you need to recognize the areas that require this filtering and network segmentation. As has been stated previously, taking a proper inventory of controls that are currently in place; understanding the business requirements; knowing local regulations, standards, and jurisdictions; and determining the industry requirements for your company all play a crucial role in determining how to design the technical and governance requirements of your security architecture.

Traffic filtering can be addressed in multiple ways within an Azure infrastructure, depending on whether you want to filter at the network, transport, or application layer. Azure Firewall, VPN Gateway, Traffic Manager, and network security groups all assist in the technical design for traffic filtering at the network and transport layers. Azure Front Door and Application Gateway used with WAFs and application security groups assist in the technical design at the application layer.

These security tools can be used to filter traffic to proper jurisdictions and geographic locations for security governance and compliance. This protects against data being accessed outside of sovereign locations and protects against potential risks of fines and sanctions against the company.

Figure 2.9 shows how a WAF policy can filter traffic and protect against bad actors and attacks coming into the application layer.

Network segmentation is another way to govern where resources and data are accessed. Resources can be placed on separate virtual networks that are virtual as well as geographically separated to protect against attacks and violations of access. Azure Firewall can be used for both network segmentation and traffic filtering along with virtual WANs, virtual networks, and network security groups. *Figure 2.11* shows how virtual subnets with network security groups can be used to segment the network between a set of virtual machines and an application with a SQL database:

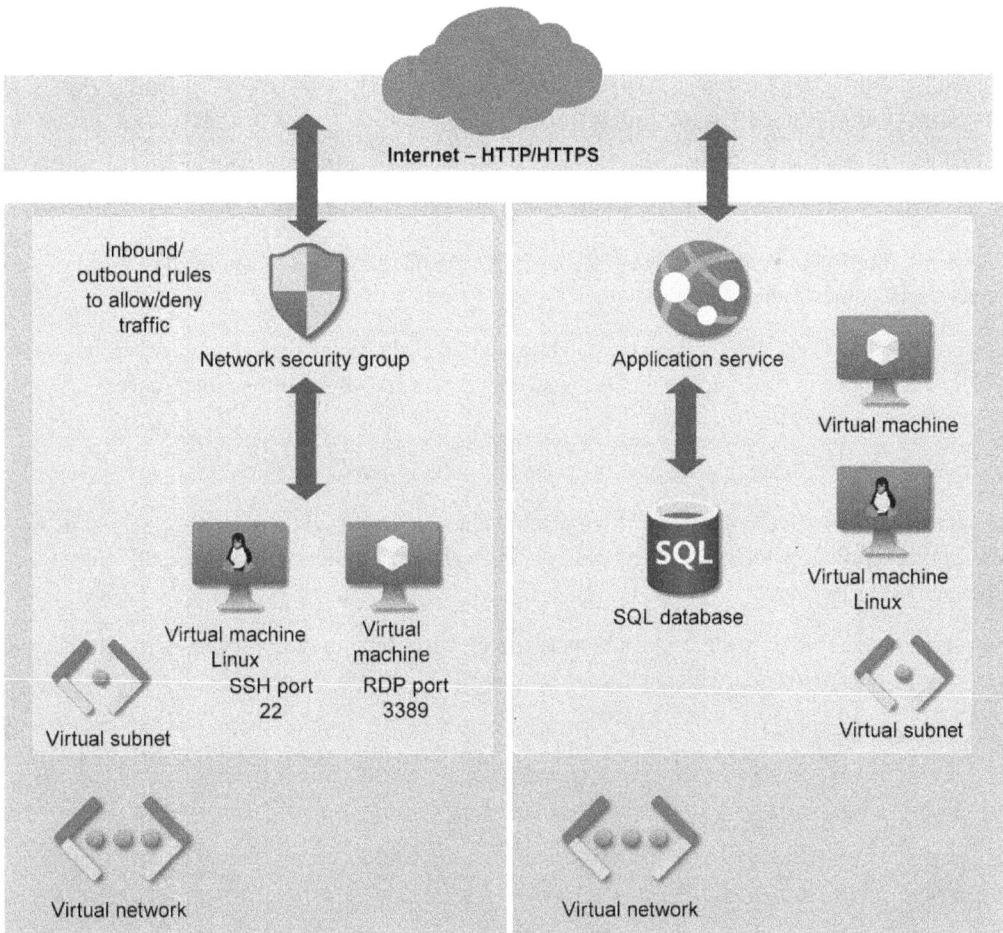

Figure 2.11: Network segmentation

Microsoft provides tools such as Network Watcher and Azure Monitor to govern the activities of network segments and resources. Microsoft Defender for Cloud and Microsoft Sentinel allow you to manage and monitor these activities.

North-South/East-West Network Traffic and Segmentation

A common technique and architectural pattern that is often used when designing a network architecture is the concept of north-south traffic and east-west traffic.

In *Figure 2.11*, north-south traffic is typically defined as traffic between the internet and cloud infrastructure and was traditionally the perimeter between your infrastructure and the rest of the world.

In a hybrid, multi-cloud environment, you may have several places where you need to monitor and control north-south traffic.

In *Figure 2.11*, east-west traffic would typically be defined as the traffic between resources that reside within your infrastructure, such as between several virtual networks and subnets, across multiple subscriptions and tenants.

This can be achieved through both hub-to-spoke and spoke-to-spoke architecture.

In a hub-to-spoke network topology, there would be a central virtual network (possibly in its own subscription) through which all traffic flows – traffic to/from the internet but also traffic from other virtual networks (spokes).

You can visualize this somewhat like a bicycle wheel, with the hub of the wheel being the central component of the network architecture, as shown in this Microsoft diagram:

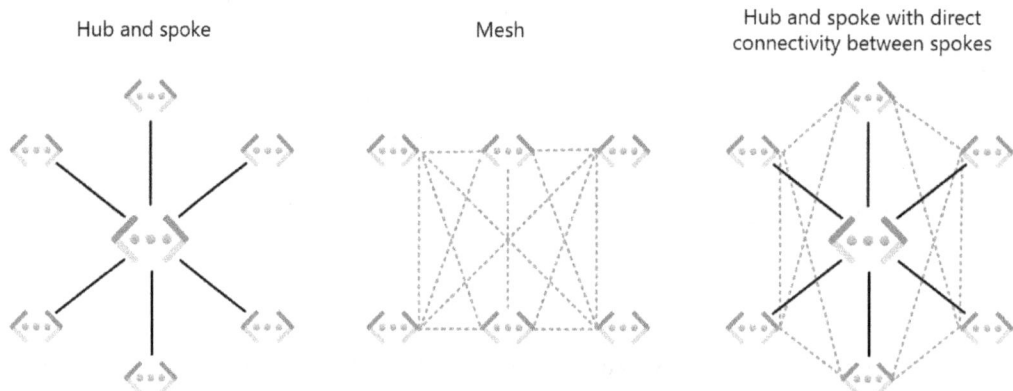

Hub and spoke   Mesh   Hub and spoke with direct
                                                         connectivity between spokes

Figure 2.12: Network topologies

There are also a number of spoke-to-spoke topologies, such as those shown in *Figure 2.12* – mesh topologies where every virtual network can talk to every other network – and also a variation of hub-to-spoke topologies, where the spokes can talk to each other independent of the hub network.

This aligns closely with the defense-in-depth strategy mentioned in *Chapter 1, Cybersecurity in the Cloud* – securing and disrupting lateral movement is critical to protecting your business from complex and ever-changing attack paths.

> **Note**
>
> You can read more about this in the Azure Architecture Center at this link: `https://learn.microsoft.com/en-us/azure/architecture/networking/guide/spoke-to-spoke-networking`.

The next section will provide a summary of the topics that were discussed in this chapter.

Summary

In this chapter, you learned about the areas of design and architecture, which are part of the overall security strategy. This included the MCRA for building that foundational architectural strategy. You also learned about the need to understand business goals to align them with your security strategy. You explored the solutions and services that can be used for a defense-in-depth security strategy. You also discovered the steps of building a resilient security architecture. This included the integration of hybrid, multi-cloud, and multi-tenant infrastructures, and technical governance for traffic filtering and network segmentation. These strategies are the foundation of the concepts of cybersecurity architecture and the zero-trust methodology.

The next chapter will discuss the strategies for designing the architecture for security operations.

Exam Readiness Drill – Chapter Review Section

Apart from mastering key concepts, strong test-taking skills under time pressure are essential for acing your certification exam. That's why developing these abilities early in your learning journey is critical.

Exam readiness drills, using the free online practice resources provided with this book, help you progressively improve your time management and test-taking skills while reinforcing the key concepts you've learned.

How to Get Started

1.   Open the link or scan the QR code at the bottom of this page.

2.   If you have unlocked the practice resources already, log in to your registered account. If you haven't, follow the instructions in *Chapter 11* and come back to this page.

3.   Once you have logged in, click the **START** button to start a quiz.

We recommend attempting a quiz multiple times till you're able to answer most of the questions correctly and well within the time limit.

You can use the following practice template to help you plan your attempts:

Working On Accuracy		
Attempt	Target	Time Limit
Attempt 1	40% or more	Till the timer runs out
Attempt 2	60% or more	Till the timer runs out
Attempt 3	75% or more	Till the timer runs out
Working On Timing		
Attempt 4	75% or more	1 minute before time limit
Attempt 5	75% or more	2 minutes before time limit
Attempt 6	75% or more	3 minutes before time limit

The above drill is just an example. Design your drills based on your own goals and make the most of the online quizzes accompanying this book.

First time accessing the online resources? 🔒

You'll need to unlock them through a one-time process. **Head to** *Chapter 11* **for instructions**.

Open Quiz	
https://packt.link/SC100_CH02	
Or scan this QR code →	

Design a Security Operations Strategy

The previous chapter discussed the design and framework for security architecture and controls based on technical and business goals. Building upon this foundation, this chapter will explore the critical aspects of designing and evaluating a strategy for security operations. Security operations encompass a wide range of activities aimed at protecting an organization's IT infrastructure and data from threats. It involves monitoring networks, detecting and responding to security incidents, and ensuring compliance with industry regulations.

This chapter will focus on the design of logging and auditing capabilities for public, hybrid, and multi-cloud environments. We will discuss how to leverage **security information and event management (SIEM)** and **security orchestration, automation, and response (SOAR)** solutions to streamline security operations and evaluate workflows and the incident management life cycle. This chapter covers the **Design security operations, identity, and compliance capabilities (25–30%)** domain of the SC-100 exam guide.

In this chapter, you are going to cover the following main topics:

- Designing a logging and auditing strategy to support security operations, including Microsoft Purview Audit
- Developing security operations to support a hybrid or multi-cloud environment
- Designing a strategy for SIEM and SOAR
- Evaluating security workflows
- Evaluating a security operations strategy for the incident management life cycle
- Evaluating a security operations strategy to share technical threat intelligence
- Leveraging artificial intelligence to supercharge security operations

Designing a Logging and Auditing Strategy to Support Security Operations, Including Microsoft Purview Audit

In the previous chapter, you learned about the key components and areas of focus to architect a security infrastructure that has proper controls in place and is resilient. Designing and architecting a secure infrastructure is important. However, attackers are constantly searching for vulnerabilities that can be exploited within the infrastructure. To contend with these vulnerabilities and threats to the infrastructure, you should be logging and auditing all activity that takes place on our infrastructure. The design of our logging and auditing strategy becomes the foundation of our company's **security operations center** (**SOC**).

Security Operations Overview

Before you go into the strategy for logging and auditing, take a moment to understand the context around security operations and an SOC. Security operations is the overall process of using information that has been gathered, and then analyzing that information for potential anomalous behavior that may indicate a threat or attack on the company resources.

An SOC comprises the tools and teams that are used to monitor, detect, and respond to threats and vulnerabilities that take place on our company infrastructure. The SOC utilizes a combination of people, processes, and technology to protect the information systems of a company. This is accomplished through continuous monitoring of systems, which involves detecting when a system, device, or identity has deviated from the standardized baseline state. Detecting these deviations promptly minimizes potential damage that can be caused by malicious activities. For the SOC to be successful, there should be a clear strategy toward what information is logged from activities and events on the cloud and hybrid infrastructure. The information that is logged will provide information that can be used for the following security operations tasks:

- Detecting threats and vulnerabilities by identifying suspicious behavior

- Investigating the threat by viewing the full attack story

- Responding to a threat through manual or automated remediation of resources that have been compromised

- Hunting for threats using logs to discover potential hidden threat activities

- Using provider threat analytics to stay informed and build context on common threats that take place globally

- Making vulnerability assessments to discover and address potential misconfigurations in resources

- Utilizing threat experts that are available from cloud and security providers for a complete understanding of attack sources

- Preventing and blocking threats through additional controls and the protection of resources

Figure 3.1 shows how integrating a security operations strategy provides a full scope to manage threats and vulnerabilities:

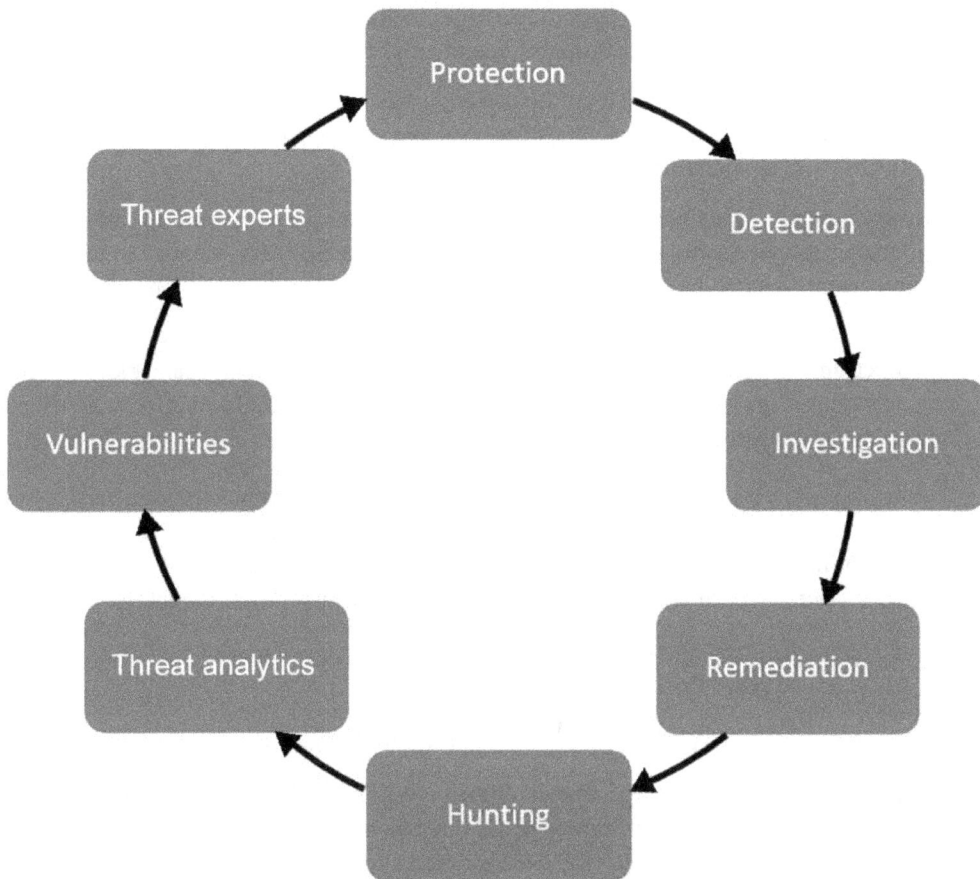

Figure 3.1: A security operations strategy to manage threats

This cycle of activities is a continuous cycle that comprises the following:

- **Detection**: Detecting threats by identifying suspicious or unusual behavior.
- **Investigation**: Investigating potential threats, as well as gathering additional context to ensure that an accurate picture of events is available to the security operations team.
- **Remediation**: Taking action to mitigate identified threats, such as disabling identities and isolating devices on the network.

- **Hunting**: Proactively searching for threats that may already be on the network that have not been detected by routine monitoring, remediating before damage can be caused to an organization.

- **Threat analytics**: Using data analysis techniques to identify, understand, and mitigate cyber threats by providing actionable insights into potential and ongoing threats.

- **Vulnerabilities**: Ensuring that an organization has a robust vulnerability management process that includes regular patching of systems and scanning for vulnerabilities throughout infrastructure and applications.

- **Threat experts**: Where necessary, engaging experts from external organizations to enhance an organization's security capability. This can be in an advisory capacity or can take a more active operational role, by complementing the existing security operations team with additional skilled resources.

- **Protection**: Employing protection mechanisms to block and defend against known attack vectors. This may involve solutions such as Microsoft Defender XDR and Microsoft Defender for Cloud, if licensed.

For security operations teams to be successful, they should be able to acknowledge and act upon an alert quickly, with automation in place that can also take the affected resources offline, decreasing the attack surface. When not responding to threats, security operations analysts should use the available tools to hunt for potential vulnerabilities and threats within a company's infrastructure.

Security operations processes and procedures should utilize the following metrics to determine their effectiveness in responding to threats:

- The **mean time to detect** is the average time it takes to discover or detect an incident.

- The **mean time to acknowledge**, or **mean time to respond**, is the time that it takes to respond to a threat. The security operations team has direct control over this metric, and if this time becomes unacceptable, the SOC may need additional tools and resources.

- The **mean time to remediate** is the effectiveness of responding to a threat and mitigating the risk to prevent it from taking place again.

- **Incidents remediated** is a measure of the manual and automated remediation that has taken place on threats and attacks. If the ratio for manual incident remediation is high, additional tools may be needed for more automated remediation before an increase in staffing is necessary.

- **Escalation between each tier** shows the effectiveness at each level of incident response. If this metric shows continuous escalation, there may be an opportunity for additional training across the incident response tiers.

Utilizing and reviewing these metrics regularly will measure the effectiveness of a company's SOC and identify areas of need that can be used for investment in additional resources, tools, and training.

> **Note**
>
> The Microsoft SOC model can be found at this link: `https://www.microsoft.com/en-us/security/blog/2019/02/21/lessons-learned-from-the-microsoft-soc-part-1-organization/`.

In the next section, you will learn about some of Microsoft's tools that can be used for security operations.

Microsoft Security Operations Tools

Microsoft provides tools within Azure and Microsoft 365 to assist in building a security operations strategy. These tools include logging services, management services, resiliency services, automation, and operations services. We will look at each of these tools in the following sections.

Azure Security Operations Services

Microsoft provides a security operations framework that companies can follow, based on their use of Microsoft infrastructure and Microsoft's unique set of controls and features, to protect cloud and hybrid infrastructures. This framework includes the **Microsoft Security Development Lifecycle**, the **Microsoft Security Response Center** program, and the overall global awareness that Microsoft has across the threat landscape. Microsoft utilizes its global partner network for its **Microsoft Intelligent Security Association (MISA)** to understand trends and global threats.

> **Note**
>
> The Microsoft Security Development Lifecycle can be found here: `https://www.microsoft.com/en-us/securityengineering/sdl/`.
>
> Microsoft Security Response Center program information can be found here: `https://www.microsoft.com/en-us/msrc`.
>
> MISA information can be found here: `https://www.microsoft.com/en-us/security/business/intelligent-security-association`.

These frameworks, programs, and associations help to create global awareness and provide Microsoft with guidance for product development of **extended detection and response (XDR)**, SIEM, and SOAR solutions.

Microsoft XDR, SIEM, and SOAR

Microsoft provides integrated cloud-native solutions that can be used for Microsoft 365, Azure, and third-party applications. These solutions include the following products:

- **Microsoft 365 Defender** is the combination of multiple Microsoft Defender solutions within Microsoft 365 for XDR and security posture management.

- **Microsoft Defender for Cloud Apps** provides a solution for managing cloud applications. This includes Microsoft 365, third-party, and on-premises applications connected and registered in Microsoft Entra. Microsoft Defender for Cloud Apps discovers applications that are accessed on a company network, and policies can be created to block activities and access to these applications. *Figure 3.2* provides a workflow of application discovery, management, and protection within Microsoft Defender for Cloud Apps that is used for security operations:

Figure 3.2: Microsoft Defender for Cloud Apps' shadow IT protection

In summary, the workflow for Microsoft Defender for Cloud Apps' shadow IT protection is a cyclical process of continual improvement that has three main phases:

- **Discover and identify**: By monitoring cloud application access intercepted by Microsoft Defender for Cloud Apps, it is possible to identify applications that were not authorized for use and evaluate their risk to the organization.

- **Evaluate and analyze**: The applications can then be evaluated against compliance policies and their usage analyzed.

- **Manage and monitor**: Finally, they can be brought under organizational management by either blocking access, allowing access to continue, and/or adding security controls, such as restricting actions that might lead to a data breach (e.g., restricting uploads).

- **Microsoft Defender for Identity** (**MDI**) and **Microsoft Entra ID Protection** protect the identities of users and devices within on-premises Active Directory and Microsoft Entra. MDI integrates the on-premises identity infrastructure with MDI sensors that monitor and identify potential risky activity, which is reported to Microsoft Defender for Cloud Apps for access policies to be executed. Microsoft Entra ID Protection monitors users and devices within the cloud for risk activities, and it allows policies to be created that enforce zero-trust verification of users and devices before they are given access to applications. *Figure 3.3* shows the architecture of MDI:

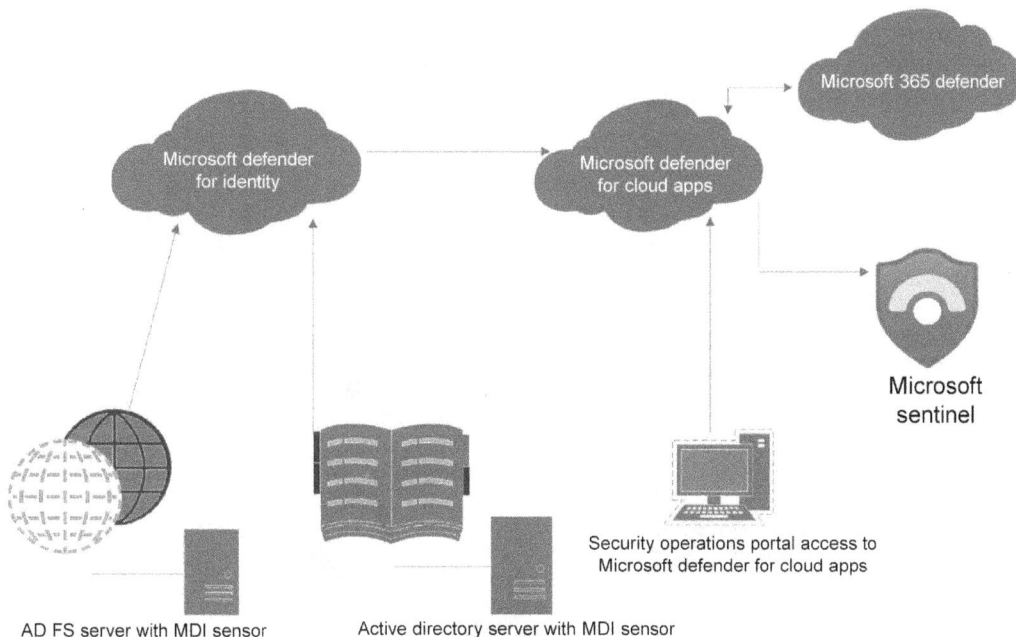

Figure 3.3: MDI

- **Microsoft Defender for Endpoint** protects managed devices. It provides threat and vulnerability management and attack surface reduction to the devices connected to Microsoft cloud services. *Figure 3.4* shows the numerous services provided with this solution:

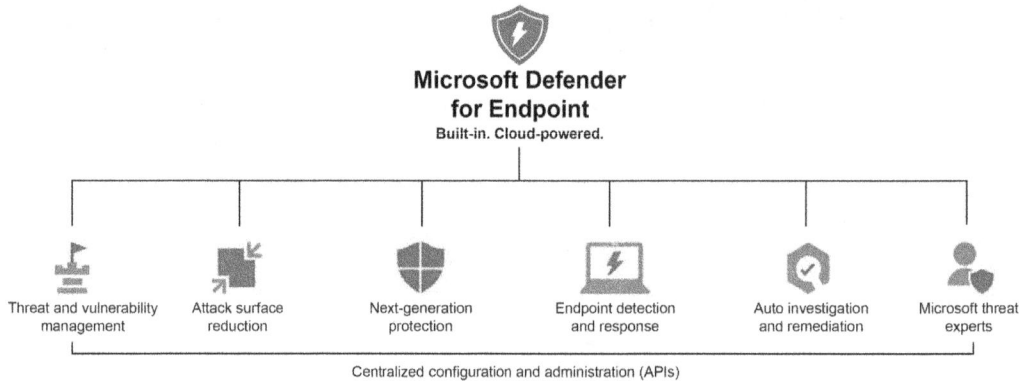

Figure 3.4: Microsoft Defender for Endpoint

- **Microsoft Defender for Office 365** protects the Office business suite, which includes Office 365, Exchange Online, SharePoint Online, and Microsoft Teams. Microsoft Defender for Office 365 detects potential malicious activity through email and collaboration communications, such as malicious links and malware, and blocks and quarantines the activity for further investigation. *Figure 3.5* shows how Microsoft Defender for Office 365 protects the Office 365 business suite:

Figure 3.5: Microsoft Defender for Office 365

- **Microsoft Defender for Cloud** provides security posture management and threat and vulnerability protection for infrastructure and platform services for Azure and non-Azure compute resources, storage accounts, applications, and databases. *Figure 3.6* shows the Microsoft Defender for Cloud security posture management process of continuously assessing and improving controls within the cloud:

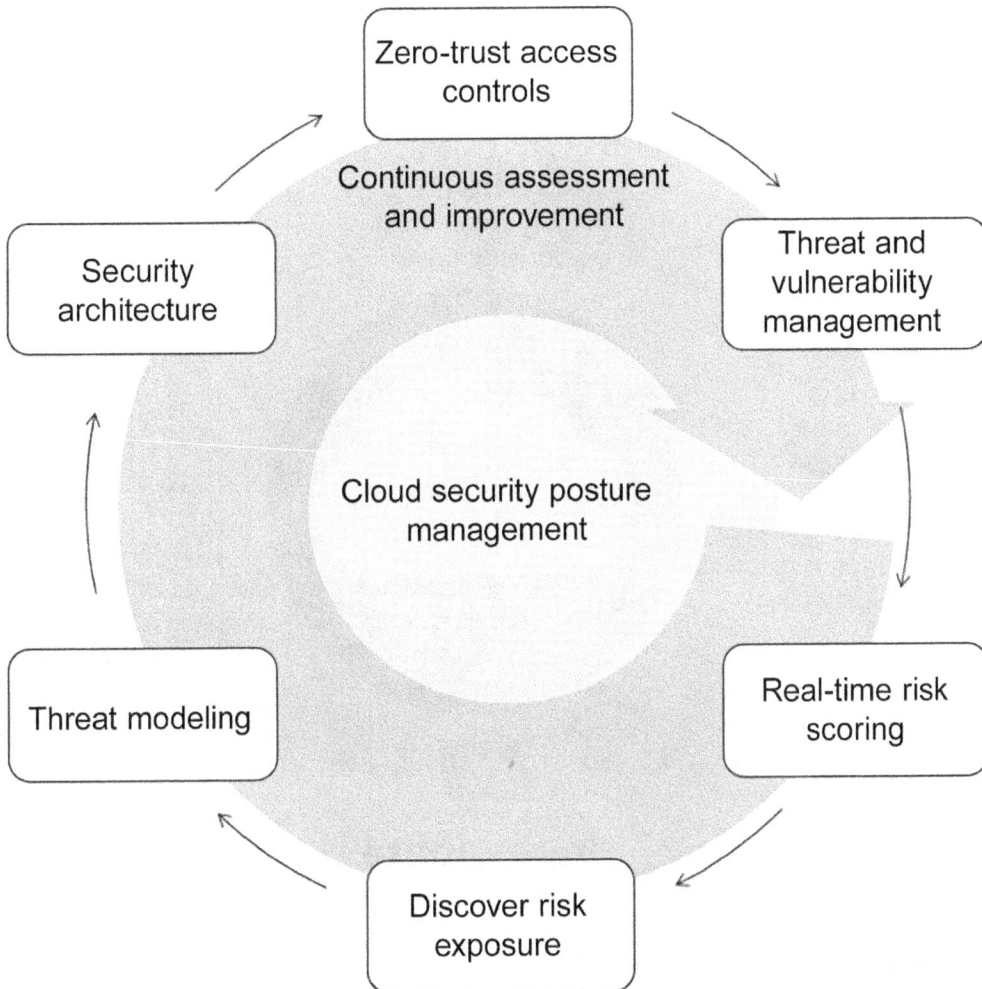

Figure 3.6: Microsoft Defender for cloud security posture management

- **Microsoft Sentinel** is a cloud-native SIEM and SOAR solution. It uses activity and event logs from Azure, Microsoft 365, AWS, Google, and other third-party security solutions to detect threats. Sentinel also hunts and investigates vulnerabilities, responding rapidly and automatically. *Figure 3.7* shows the complete workflow of Microsoft Sentinel:

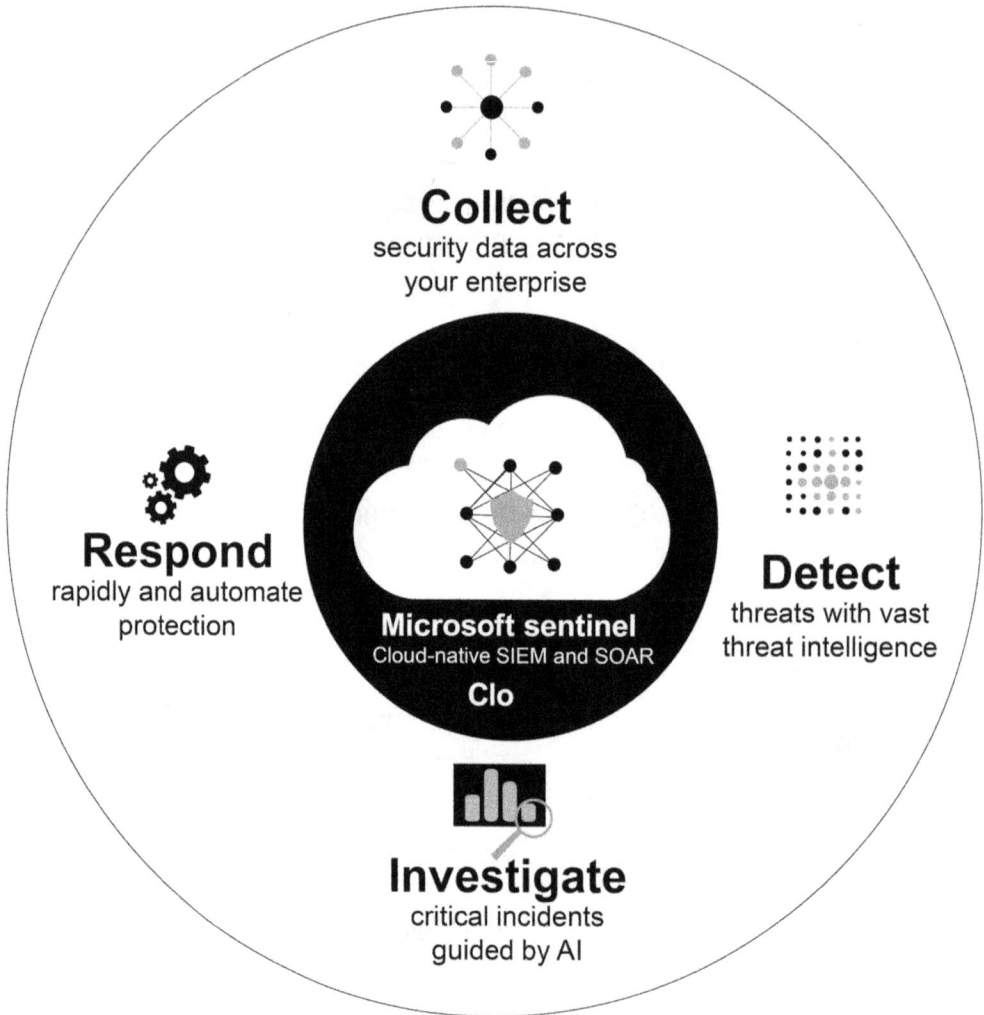

Figure 3.7: Microsoft Sentinel SIEM and SOAR

Microsoft Defender XDR solutions from Microsoft 365 Defender and Microsoft Defender for Cloud tie into the Microsoft Sentinel SIEM and SOAR capabilities, as shown in *Figure 3.8*:

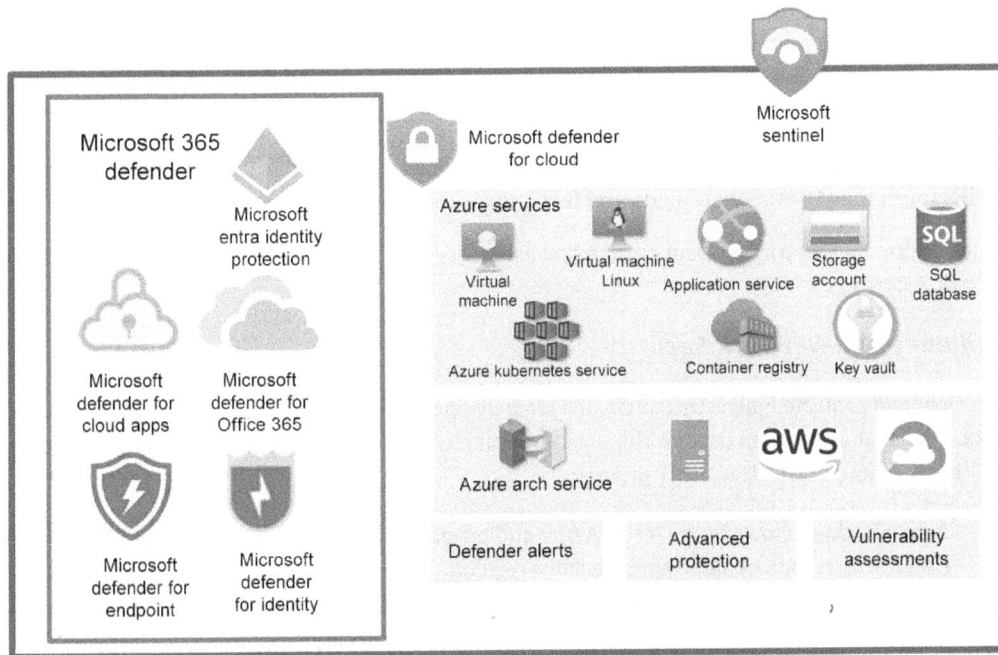

Figure 3.8: Microsoft 365 Defender and Microsoft Sentinel

These solutions allow companies to access capabilities that can be used to build or expand their security operations. Being able to discover and gather resources, applications, and identities and collect activity and events provides the tools for a security operations team to analyze and respond to threats and vulnerabilities.

Microsoft automation services create operations efficiencies through workflows that limit the attack surface when an attacker compromises a resource.

Automation Tools

Azure Automation is used to eliminate common manual tasks that are repeatable. This automation includes updates and security patches on virtual machines and other compute resources. Azure Automation allows these tasks to be scheduled and performed at regular intervals, ensuring that resources are up to date and avoiding common vulnerabilities due to delays in updates and patches.

Logic Apps is integrated into Microsoft Defender for Cloud and Microsoft Sentinel to extend capabilities and create workflows, based on alert triggers that execute a notification or initiate an activity on resources. These activities may include shutting down a virtual machine that is potentially compromised and under attack to cut off the attacker while investigating the attack that has taken place. Logic Apps uses a graphical design interface to create these workflows, and then it connects them to Microsoft Defender for Cloud and Microsoft Sentinel.

The automation of common tasks and alert response workflows increases the mean time to acknowledge and the mean time to remediate alerts and incidents within security operations.

Next, let's discuss how a cybersecurity architect assists security operations with proper resiliency and network security.

Resiliency and Network Security

A resilient and protected infrastructure helps security operations protect resources and get resources back online after an incident occurs. This can be accomplished through Azure services such as Backup, Site Recovery, and network security groups:

- **Azure Backup** allows resources in Azure and on-premises to be backed up in the Azure cloud. Backing up resources and testing backups regularly can protect a company against attacks such as ransomware or malware. When a company has a valid backup of an affected resource, that resource can be restored to the point before they were infected.

- **Azure Site Recovery** (**ASR**) is a cloud-based solution that provides business continuity and disaster recovery capabilities. ASR creates a failover site to another Azure region to allow resources in another region or on-premises. ASR is also a cost-effective solution for a recovery site without renting space and purchasing duplicate hardware.

- **Azure network security groups** are used to provide inbound and outbound rules to filter traffic on ports and IP addresses. Network security groups are associated with virtual network subnets or virtual network interfaces of virtual machines, allowing or denying traffic. They can be grouped into ASGs to reduce complexity when handling traffic filtering across applications. Intelligently designing these rules can assist in protecting network resources in Azure against potential threats.

The activity of these services is integrated into the following monitoring and management services within Azure.

Azure Monitoring and Management Services

The IT and security operations of a company can only be successful if the resources that they use are properly monitored and managed throughout the cloud and hybrid infrastructure. This requires proper logging of events and activities across resources, as well as having the tools in place to take this information and provide insight into unusual activities. Microsoft has tools that can capture this information and provide these insights. These monitoring tools are as follows:

- **Azure Monitor** is a service connected to Azure resources to capture and log events and activities, as well as performance metrics. This information can then be used for security operations for insights and threat analysis, using Log Analytics.

- **Azure Arc** is used to capture similar information as Azure Monitor for hybrid resources. Azure Arc bridges the on-premises, non-Azure resources with Azure architecture for consistent monitoring and managing resources.

- **Log Analytics** uses the event and activity log data from Azure Monitor and Azure Arc to provide the valuable insights needed to manage the infrastructure resources. Queries can be run against the log data to identify suspicious activity within an environment.

- **Network Watcher** monitors and logs traffic within the Azure virtual network and on-premises networks where agents have been installed. This data can be used to analyze potential threat activity and network vulnerabilities.

The log data from these services is the foundation of designing a logging and auditing strategy for security operations.

Logging and Auditing for Threat and Vulnerability Detection

The success of security operations depends on logging and monitoring all resources within the cloud and hybrid infrastructure. This strategy relies on deploying logging agents and monitoring the data that is collected. When an alert is created, it must provide a notification for escalation within security operations. The security operations team must take that alert and respond to the incident. Information and details from logs should be audited regularly. Reviewing and investigating these logs could uncover potential vulnerabilities and threats before they are exploited in attacks on company resources.

Log data can be provided from Azure activity logs, data plane and diagnostic logs, and processed events from alert events, within solutions such as Microsoft Defender for Cloud and Microsoft 365 Defender. This log data provides insights into applications, allowing you to troubleshoot potential past incidents and prevent future events from taking place by putting new security controls in place. Some prevention activities can be automated with Azure Automation or Logic Apps workflows within Log Analytics.

Microsoft Purview Audit

Microsoft Purview Audit enhances your security operations by providing an integrated solution for responding to security incidents, forensic investigations, and compliance obligations. You can capture, record, and retain thousands of user and admin activities across various Microsoft services in a unified audit log.

The key benefits include the following:

- **Comprehensive event logging**: Tracking and searching for a wide range of audited activities
- **Custom retention policies**: Retaining audit records based on specific criteria for up to 10 years
- **High-bandwidth access**: Quickly accessing audit data through the Office 365 Management Activity API
- **Intelligent insights**: Gaining valuable insights to support your investigations and compliance efforts

By leveraging these features, you can effectively monitor and investigate potential security breaches, ensuring that your organization remains compliant and secure.

Comparing Audit (Standard) and Audit (Premium)

Purview Audit is available in two different SKUs – Standard and Premium.

You can create customized audit log retention policies to retain audit records for up to 1 year, or up to 10 years with the required add-on license. Policies can be based on the service, specific activities, or user involved.

By default, audit records for Microsoft Entra ID, Exchange, OneDrive, and SharePoint are retained for one year, while other activities are retained for 180 days (about 6 months). You can extend these periods using retention policies.

Audit (Premium) offers intelligent insights to aid forensic and compliance investigations, providing visibility into events such as mail access and user searches in Exchange Online and SharePoint Online.

Additionally, Audit (Premium) grants higher bandwidth to the Office 365 Management Activity API, dynamically increasing request limits based on your organization's seat count and licensing, effectively doubling the bandwidth compared to Audit (Standard).

Capability	Audit (Standard)	Audit (Premium)
Enabled by default	Yes	Yes
Thousands of searchable audit events	Yes	Yes
Audit search tool in the Microsoft Purview portal and compliance portal	Yes	Yes

Capability	Audit (Standard)	Audit (Premium)
Search-UnifiedAuditLog cmdlet	Yes	Yes
Export audit records to a CSV file	Yes	Yes
Access to audit logs via the Office 365 Management Activity API	Yes	Yes
180-day audit log retention	Yes	Yes
1-year audit log retention		Yes
10-year audit log retention		Yes
Audit log retention policies		Yes
Intelligent insights		Yes

Table 3.1: Audit capabilities: Standard vs. Premium

Scenarios Enabled by Purview Audit

Some common use cases/scenarios for Purview Audit from a security operations perspective are as follows:

- Finding the IP address of a computer used to access a compromised account
- Determining who set up email forwarding for a mailbox
- Determining whether a user deleted email items in their mailbox
- Determining whether a user created an inbox rule
- Investigating why there was a successful sign-in by a user outside your organization
- Searching for mailbox activities performed by users with non-E5 licenses
- Searching for mailbox activities performed by delegate users
- Auditing activities in Microsoft Teams, SharePoint, and OneDrive to provide deeper insights into user activities across multiple platforms, enriching and adding context to a security incident and aiding the SOC analyst in their investigation

> **Note**
>
> You can find detailed pricing information for Microsoft Purview in the **M365EnterprisePlans** document, available at this link: `https://aka.ms/M365EnterprisePlans`.

This section provided an overview of security operations and the tools that Microsoft provides to log, monitor, manage, and audit resources.

In many companies, infrastructure may include Microsoft 365, Azure, on-premises, and other cloud provider resources. This creates complexity in a strategy for security operations. In the next section, you will learn how to develop this hybrid and multi-cloud security operation environment.

Developing Security Operations to Support a Hybrid or Multi-Cloud Environment

Most enterprises have not become fully cloud-native, meaning that their entire infrastructure utilizes cloud services. A cloud-native company would have all business applications that use Microsoft 365, a customer relationship manager with Dynamics 365, databases on Azure SQL Database, and custom applications running in Azure. However, this is not the reality. These companies may utilize Microsoft 365, have databases and SharePoint servers in their on-premises data centers, and even store data in **Amazon Web Services (AWS)** S3 buckets or on the **Google Cloud Platform (GCP)** BigQuery platform.

Hybrid cloud companies are companies that use cloud services, such as Azure, along with on-premises infrastructure. Adding the use of more than one cloud provider, such as Azure and AWS or GCP, to the mix has led to the widespread adoption of the term "multi-cloud" in the industry. *Figure 3.9* provides a diagram of this hybrid and multi-cloud architecture for reference:

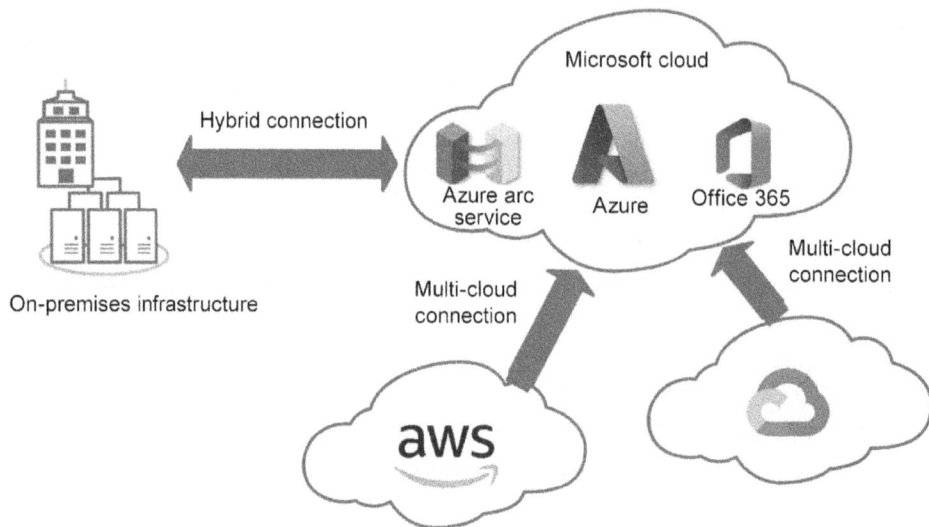

Figure 3.9: Hybrid and multi-cloud architecture

> **Note**
> Many companies would be defined as multi-cloud companies if they utilized more than one SaaS solution. They may have a private data center but use Microsoft 365, Workday, and Salesforce. Workday and Salesforce are both cloud-native SaaS solutions, so the company should be classified as multi-cloud.

The objective of a cybersecurity architect is to design a strategy and solution that can bring these on-premises and cloud-native solutions together for security operations to monitor, manage, and investigate threats and vulnerabilities on a single platform. Proprietary tools and technologies across cloud providers make it difficult to manage a hybrid architecture, but providing some level of baseline consistency in how resources are deployed and governed can be achieved with Microsoft tools.

Microsoft Defender for Cloud can monitor and manage the cloud security posture of Azure resources, along with joining AWS accounts and GCP projects for unified log, event, and activity analysis for compute and storage resources. Azure Arc also provides a connection bridge for on-premises compute and databases to be monitored and managed within an Azure environment. Arc can also be used with cloud providers outside of AWS and GCP to connect virtual machines to Azure. Microsoft Defender for Cloud Apps provides tools to discover, monitor, and manage access to and activity on third-party, cloud-native SaaS applications.

Figure 3.10 shows a simplified version of security operations across a company's various points of data activity, as described in the **Microsoft Cybersecurity Reference Architecture (MCRA)**. The official diagram can be found in the 58th slide at this URL: `https://github.com/MicrosoftDocs/security/blob/main/Downloads/mcra-december-2023.pptx?raw=true`.

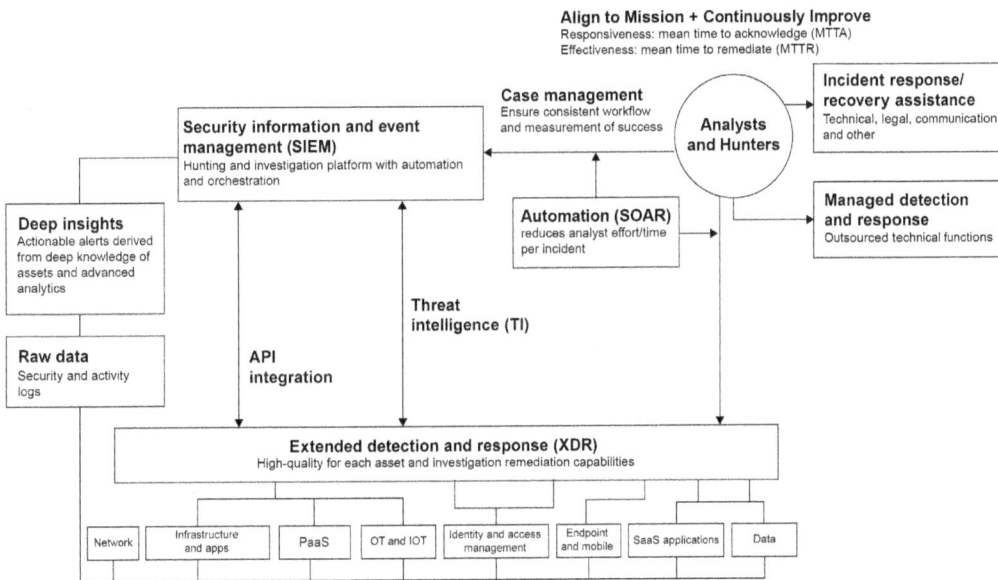

Figure 3.10: Security operations technology usage

The raw data, referenced in *Figure 3.10*, comes from numerous hybrid and multi-cloud sources that need to be accounted for within a security operations strategy, to monitor and manage threats and vulnerabilities.

The goal of a cybersecurity architect is to recognize these data points and design an architecture that facilitates ease of access to the security activity and event logs from these resources.

The next section will discuss how an SIEM and SOAR solution can be used to impact this design, as well as facilitate access to investigating and responding to activity and events from these data sources.

Designing a strategy for SIEM and SOAR

As stated in the previous section, an important aspect of the security operations strategy is the ability to create an architecture that utilizes tools for the SOC team to hunt and investigate activity and event log data from multiple sources. SIEM and SOAR solutions can facilitate this capability. Let's define the two for clarity:

- **SIEM**: An SIEM solution is usually deployed within a security operations center that gathers logs and events from various appliances and software within an IT infrastructure. An SIEM solution then analyzes the logs and events for potential threats by searching for behavior that is not typical of best practices or may be seen as anomalous or atypical. Without an SIEM, security operations personnel would need to review each log and event file manually. Since there are thousands of log and event files within companies, this option has the potential for mistakes, as fatigue becomes an issue when analyzing and identifying log changes. An SIEM identifies the logs and events that could be a threat; then, security personnel can investigate these potential threats. This decreases the time to recognize a threat or vulnerability, allowing the security operations team to be more efficient and effective in their investigations.

- **SOAR**: A SOAR solution is a complementary solution to an SIEM. By initiating a workflow, SOAR solutions can add automation to the response of potential events identified as threats in the log files. An example of this would be an activity log from a device accessed from a location that has been flagged as a threat. SOAR can initiate a workflow to take that device offline and send an alert to the security operations response team to investigate.

Microsoft has these capabilities with Microsoft Sentinel. *Figure 3.11* shows the architecture of Microsoft Sentinel:

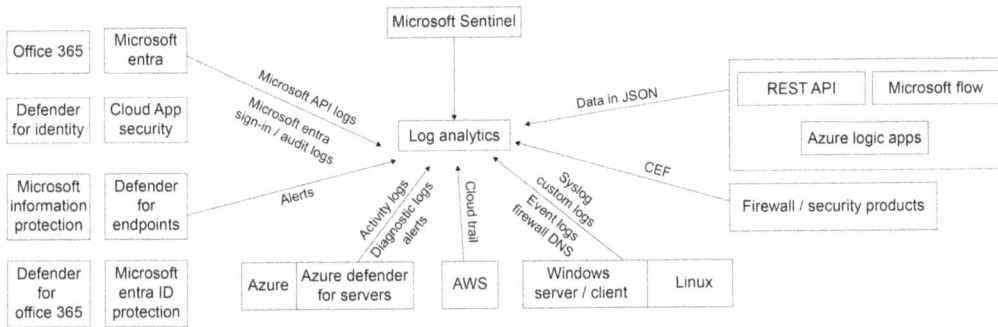

Figure 3.11: Microsoft Sentinel architecture

Microsoft Sentinel can gather and coordinate information from data sources within Microsoft 365, Azure, Microsoft Entra, AWS, Windows and Linux devices, third-party security products, and API connections from applications. This information can be used with the insights from Log Analytics to run queries in **Kusto Query Language** (**KQL**) to help identify potential anomalies and threats.

In addition, Microsoft Sentinel has automated queries and tasks that run in the background and are modeled with Microsoft machine learning tools, identifying anomalous behavior on identities and data within an environment.

An intelligently designed SIEM and SOAR solution will increase the efficiency of security operations.

> **Note**
> For additional information on Microsoft's best practices for SIEM and SOAR, review the
> MCRA: `https://learn.microsoft.com/en-us/security/adoption/mcra`.

A cloud-native SIEM/SOAR solution such as Microsoft Sentinel can increase the scalability of capabilities and the speed of response to incidents across hybrid and multi-cloud environments. The automated investigation and integration with the MITRE ATT&CK framework allow security operations to prioritize an investigation by identifying the most serious vulnerabilities. The MITRE ATT&CK framework was discussed in *Chapter 1, Cybersecurity in the Cloud*. The integration of Microsoft Sentinel with Microsoft's XDR solutions in Microsoft 365 Defender and Microsoft Defender for Cloud can increase the accuracy of the hunting and investigating process and reduce the number of false positives.

Now that you understand the requirements and tools to design a security operations strategy, you can evaluate and review how these tools impact the response to incidents and security workflows.

In the next section, you will be reviewing the security strategies for incident management and evaluate security workflows.

Evaluating Security Workflows

The impact that security operations have on a company is based on the ability to recognize and respond to incidents. The most efficient security operations teams and tools will also continuously hunt for vulnerabilities and threats, enabling new controls before a threat becomes an incident. However, incidents are going to happen within any company. Let's define what an incident is and how security operations respond to this type of threat.

Security Strategies for Incident Management and Response

A security incident is a confirmed breach in a company environment that could lead to access, loss, or disclosure of sensitive company information, such as customer or personal data. NIST 800-61 R2 states that a "*security incident is a violation or imminent threat of violation of computer security policies, acceptable use policies, or standard security practices.*"

A company should have a strategy and process for responding to incidents, including identifying an incident response team. These incident response teams include members of the SOC, as well as other teams within an IT organization and management stakeholders. In addition, public relations and communications teams should be part of the incident response to notify customers and partners of potential information breaches.

These cross-functioning teams will work together to eliminate the threat of replicating in an environment, investigate the source, and remove the threat from affected systems. These teams will then work together on a response communication and explain the controls that have been put in place to mitigate the threat from taking place in the future.

This process, as defined by the NIST 800-61r2 approach, can be described in the following steps:

1. Preparation refers to the actions that a company needs to take. This includes identifying the tools and team required for the processes, competencies, and readiness to respond to an incident.

2. Detection and analysis refer to the ability to detect incidents and analyze them to confirm that they are a threat, not a false positive. SIEM and SOAR solutions assist in this area.

3. Containment, eradication, and recovery are the actions that the incident response team must take to identify the resource that has been affected, cutting it off from production on time. After the threat has been contained, the incident response team can work to remove the threat from the resource and put the resource back into production. In some cases, the resource may be compromised to the point that a system backup from another recovery point before the attack may be required.

4. Post-incident activity includes the communication that needs to occur between internal and external stakeholders and post-mortem analysis on the efficiency and effectiveness of recovery. This activity will also encompass the controls that were put in place to remediate and mitigate a repeat of the same incident.

> **Note**
>
> More information on this approach can be found at this link: `https://nvlpubs.nist.gov/nistpubs/SpecialPublications/NIST.SP.800-61r2.pdf`.

Any incident response process requires this post-incident activity to determine areas for improvement. These activities, along with continuous hunting for threats and vulnerabilities, contribute to an efficient and effective level of security operations. *Figure 3.12* shows the workflow of the continuous improvement process. It is a simplified version of an official Microsoft diagram from the MCRA, which can be viewed in the 55th slide at the following URL: `https://github.com/MicrosoftDocs/security/blob/main/Downloads/mcra-december-2023.pptx?raw=true`.

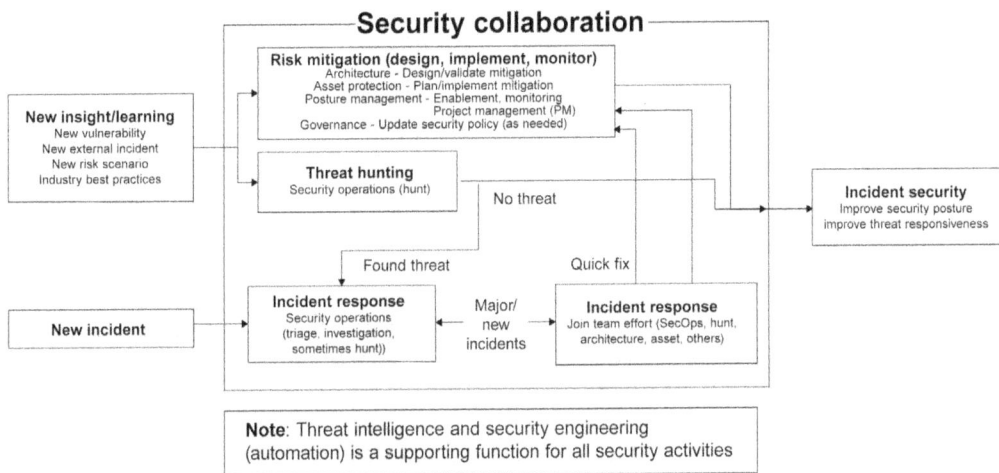

Figure 3.12: The security operations continuous improvement process

Evaluating and improving the tools and techniques within security operations will mitigate risks from becoming attacks. The cybersecurity architect should work with security operations and cross-functional teams to review the effectiveness of a company's security incident response.

Security Workflows

The response to an incident and the workflow within a company is not the same for all incidents. The recommendation is that you coordinate workflows with automation within the processes and procedures. An example of this is creating a Logic Apps notification workflow for the different teams required to respond to certain attacks and incidents. The same teams are not going to be involved in an identity-based attack as the teams involved in a network port breach.

When evaluating and determining the workflow for a response, the following should be accounted for when evaluating security operations:

- Identify and determine the attack scope with an understanding that attackers may use different and multiple mechanisms to execute an attack.

- Identify the objective and target of the attack to understand where an attacker exploits system vulnerabilities.

- Stay focused on sensitive and business-critical data to minimize the potential impact on customers and personal data.

- Be clear in the process and coordination of the response team, with identified roles and responsibilities for each team member.

- Maintain the business perspective on the impact of an incident. This perspective relates to the scope of the attack, where you should put yourself in the mind of the attacker and understand what the goal is for business operations.

- Break up the response scope into executable tasks by working on operations that can be executed within less than a day.

- Maintain a focus on the short term before thinking about the long term. Security controls can be put into place later to protect against long-term vulnerabilities. The response and remediation of the incident should be the primary focus.

- Keep a clear plan for teams and limit the scope with documented and clear processes and procedures for response and the flow of information throughout the team.

- Outline the ownership of stakeholders and the hierarchy for quick and decisive decisions.

- Continue to communicate to stakeholders with frequent updates, with expectations on the activities that take place in response to the attack.

You should have these processes and procedures for incident response workflows documented and shared with all stakeholders who are part of the team. This will maintain a clear path to execute a response to an attack effectively and efficiently. As stated previously, where available, build automation into the incident response workflow with Microsoft technologies. Some of these technologies have been previously discussed, such as Azure Logic Apps, Microsoft Defender for Cloud, and Microsoft Sentinel. Microsoft Graph Security is also a helpful API that helps unify Microsoft security tools with a third-party security solution ecosystem.

> **Note**
>
> More information on the Microsoft Graph Security API can be found at this link: https://docs.microsoft.com/graph/api/resources/security-api-overview?view=graph-rest-1.0.

A clearly defined workflow for security response is the foundation for effectively and efficiently responding to an incident. These processes and workflows should be continually evaluated through post-mortem activities and lessons-learned discussions, allowing you to understand the tiered approach to escalated attack response. This creates an incident management life cycle for response and a tiered escalation for an attack.

The following section will teach you how to evaluate the incident management life cycle.

Evaluating a Security Operations Strategy for the Incident Management Life Cycle

The life cycle of managing an incident response can be viewed as a tiered approach. The more prepared security operations are to identify and respond to threats, the lower the impact on a business from a financial, personal, and reputational perspective.

Figure 3.13 shows the tiers of the various security operations functions:

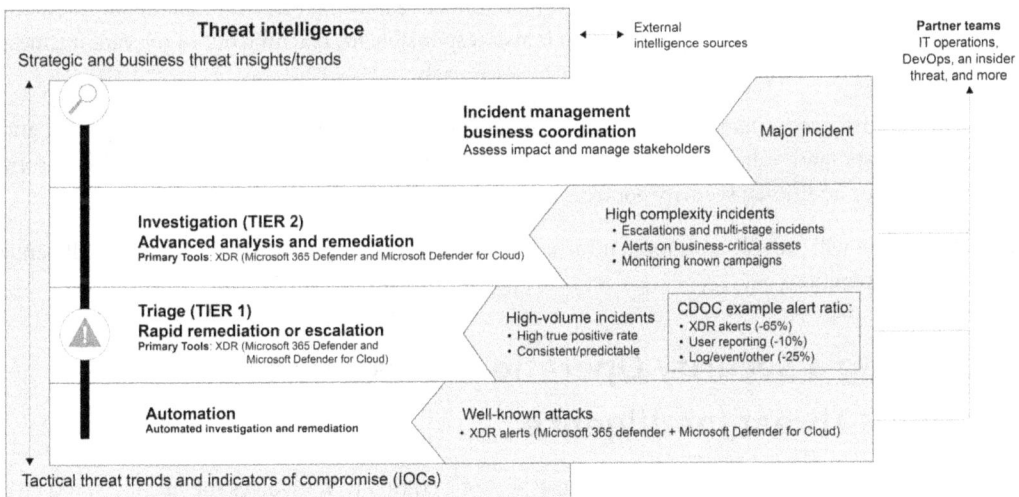

Figure 3.13: The security operations functional tiers

As shown in *Figure 3.13*, automation in the workflow is important to managing the efficiency of a security operations team. Implementing automated responses for common attacks, such as phishing URLs, SQL injections, brute-force attacks, or port scans, enhances efficiency in the security operations team. An example would be a **web application firewall (WAF)**, which protects against SQL injections, or Microsoft Defender for Office 365, which protects against phishing. Another example is **multi-factor authentication (MFA)**, which protects against brute-force identity attacks, and **just-in-time (JIT)** virtual machine access, which protects against port scans.

This automation protects from attacks and will also log them in solutions such as Microsoft Defender for Cloud, allowing additional triage of them. Some attacks may break through the controls and protection that are in place. This would require additional escalation to tier 1 for triage, which involves reviewing and remediating a high volume of incidences of these common attacks. Triage is the process that solves the most common incident types and resolves them within the team. More complex incidents or incidents that have not been seen and resolved before should be escalated to the next tier, which is tier 2.

More complex attacks that may impact multiple systems and assets will require escalation for additional investigation. This tier 2 team begins to review and analyze business-critical assessments and ongoing attack campaigns that are known through global threat intelligence information. The tier 2 team is focused on remediating the active attack while also monitoring to identify new attacks.

The hunting team at tier 3 is an escalation point for tier 2 when responding to complex multi-stage attacks, but they also are a proactive group that handles advanced forensics and detects potential threats and vulnerabilities before they are exploited. Effective hunting can eliminate major incidents before they become attacks. The hunting team is also responsible for reaching out to relevant business stakeholders to create a task force if there is a major breach.

Within the security ecosystem, cybersecurity professionals within a company, partner companies, and industry analysts share valuable insights that can help protect the company, as well as build a strong business case for additional security controls and tools.

The next section will provide additional information on how you can integrate this threat intelligence into security operations.

Evaluating a Security Operations Strategy to Share Technical Threat Intelligence

Security operations is a continuously evolving aspect of a company. It involves the need to stay a step or two ahead of the constant addition of new and more complex attacks, as well as the omnipresent threats, such as phishing, malicious URLs, and brute-force attacks. Constant learning and sharing of technical threat intelligence is necessary for effective security operations.

Threat intelligence gathers data on a potential attack that assists you in defining the scope and objective of the threat. This is defined in the alert with characteristics of the attack, the potential location latitude and IP address of the attack, and the resource that is the target of the attack. Microsoft Sentinel utilizes Microsoft's network of global threat intelligence professionals and partners. This threat intelligence is essential to hunting and investigating threats and gathering information across data sources. Microsoft 365 Defender XDR provides similar threat intelligence gathering, for hunting on Microsoft 365 and decreasing the attack surface on devices with Microsoft Defender for Endpoint. Microsoft Defender for Cloud utilizes this threat intelligence to alert you about potential attacker identities and the MITRE ATT&CK objectives. Defender for IoT provides tools for connected devices that ingest raw diagnostic data, protecting it at the network level with SOC tools.

> **Note**
>
> Information on Microsoft Threat Intelligence can be found at this link: `https://www.microsoft.com/en-us/security`. Microsoft also publishes information on the global threat landscape on the Microsoft Security Intelligence page, which can be found here: `https://www.microsoft.com/en-us/wdsi/threats`.

The ability of security operations to utilize the information provided by Microsoft, as well as the ecosystem of security partner tools, provides an effective and efficient way to understand common threats and vulnerabilities before they become incidents that your company needs to respond to. The cybersecurity architect will work with security operations teams to continuously evaluate the tools and controls that are used to hunt and recognize threats before they are exploited.

Leveraging Artificial Intelligence to Enhance Security Operations

> **Note**
>
> At the time of writing (July 22, 2024), **Microsoft Copilot for Security** is not part of the SC-100 Exam curriculum. It is reasonable to expect that it may become part of the exam curriculum in the next update, having gone into **general availability (GA)** on April 1, 2024. Although Microsoft Copilot for Security is not currently part of the test, you should be familiar with it for future curriculum updates. You may also find it useful to enhance your security operations capabilities.

There has been a significant increase in the power, accuracy, popularity, and availability of **generative AI (GenAI)** for both malicious and non-malicious use.

> **Note**
>
> You can read more about that trend in the report at this link: `https://www.mckinsey.com/capabilities/quantumblack/our-insights/the-state-of-ai`.

With the increase in usage of GenAI, aligned with the ever-increasing scale and complexity of cloud infrastructure, applications, and threats, as a cybersecurity architect, you should be aware that AI helps to accelerate security operations.

Microsoft Copilot for Security

Microsoft Copilot for Security is a virtual AI assistant that you can ask questions, in natural language (known as a prompt).

Based on the text prompt, Microsoft Copilot for Security can do the following:

- Summarize incidents in Microsoft Sentinel
- Enrich incidents with relevant additional context relating to identities, devices, **indicators of compromise (IoCs)**, and many more information sources
- Analyze workloads and identify attack paths
- Analyze scripts for signs of malicious behavior

It is integrated with several products within the Microsoft Azure and Microsoft 365 stack, such as Microsoft Sentinel, Microsoft Defender, Microsoft Entra, and Microsoft Purview.

> **Note**
>
> You can read more about the integrations at this link: `https://www.microsoft.com/en-us/security/business/ai-machine-learning/microsoft-copilot-security#Integrations`.

This can provide several benefits to your security operations, such as the following:

- Reduce missed signals of potentially malicious behavior
- Reduce the mean time to respond to incidents
- Increase the knowledge and efficiency of your SOC

> **Note**
>
> You can find out more about Microsoft Copilot for Security at this link: `https://www.microsoft.com/en-us/security/business/ai-machine-learning/microsoft-copilot-security`.
>
> You can also undertake further training on Microsoft Copilot for Security via **Microsoft Learn** at this link: `https://learn.microsoft.com/en-us/copilot/security`.

The next section will provide a summary of the topics that were discussed in this chapter.

Summary

In this chapter, you covered the areas of design and evaluation that go into an overall security operations strategy. This included the tools and teams that are assembled for an SOC and the tools that are available from Microsoft to monitor, manage, and respond to threats within the security operations process.

The use of SIEM, SOAR, and XDR tools to automate responses to threats and hunt for potential threats is the foundation of security operations. Utilizing these tools, with a strong security operations workflow for incident response, becomes the gauge to evaluate an effective and efficient security operations and incident response team. The cybersecurity architect will work closely with security operations to design and evaluate the tools, techniques, processes, and procedures for incident response, effective tools, and automation.

You were also introduced to Microsoft Copilot for Security, including its advantages in increasing the efficiency of your security operations.

In the next chapter, you will learn about the design and architecture of an effective identity security strategy.

Exam Readiness Drill – Chapter Review Section

Apart from mastering key concepts, strong test-taking skills under time pressure are essential for acing your certification exam. That's why developing these abilities early in your learning journey is critical.

Exam readiness drills, using the free online practice resources provided with this book, help you progressively improve your time management and test-taking skills while reinforcing the key concepts you've learned.

How to Get Started

1. Open the link or scan the QR code at the bottom of this page.

2. If you have unlocked the practice resources already, log in to your registered account. If you haven't, follow the instructions in *Chapter 11* and come back to this page.

3. Once you have logged in, click the **START** button to start a quiz.

We recommend attempting a quiz multiple times till you're able to answer most of the questions correctly and well within the time limit.

You can use the following practice template to help you plan your attempts:

Working On Accuracy		
Attempt	Target	Time Limit
Attempt 1	40% or more	Till the timer runs out
Attempt 2	60% or more	Till the timer runs out
Attempt 3	75% or more	Till the timer runs out
Working On Timing		
Attempt 4	75% or more	1 minute before time limit
Attempt 5	75% or more	2 minutes before time limit
Attempt 6	75% or more	3 minutes before time limit

The above drill is just an example. Design your drills based on your own goals and make the most of the online quizzes accompanying this book.

First time accessing the online resources? 🔓
You'll need to unlock them through a one-time process. **Head to *Chapter 11* for instructions.**

Open Quiz	
https://packt.link/SC100_CH03	
Or scan this QR code →	

4
Design an Identity Security Strategy

The previous chapter discussed the design of an identity security strategy and how to evaluate it for security operations while utilizing the concept of zero trust. This chapter will focus on designing an identity security strategy for cloud-native, hybrid, and multi-cloud identity and access management infrastructures. It will cover the design criteria and recommendations of a zero trust strategy for internal tenants, external customers and partners, and hybrid architectures.

This chapter will help with your preparation for and understanding of some of the key domains of the SC 100 Exam Guide, specifically, **Design security operations, identity, and compliance capabilities (25–30%)** and **Design solutions for identity and access management**.

In this chapter, you are going to cover the following main topics:

- Zero trust for identity and access management
- Designing a strategy for access to cloud resources
- Recommending an identity store (tenants, B2B, B2C)
- Recommending an authentication and authorization strategy
- Designing a strategy for **Conditional Access (CA)**
- Designing a strategy for **continuous access evaluation (CAE)**
- Designing a strategy for role assignment and delegation
- Designing a security strategy for privileged role access
- Designing a security strategy for privileged activities
- Case study – designing a zero-trust architecture

Once you have completed this chapter, you will be able to design a strategy for identity and access management.

Zero Trust for Identity and Access Management

In the previous chapter, you learned about the design of an identity security strategy and how to evaluate a strategy for security operations. Managing and monitoring potential threats and vulnerabilities with security operations is aided by a zero-trust strategy. This is increasingly true when determining a strategy for recommendations to design a secure identity and access architecture. Let us review some of the foundational elements of zero trust and how they relate to securing identities.

Zero Trust is an integrated approach to securing access with adaptive controls and continuous verification across your entire digital estate. Everything from the user's identity to the application's hosting environment verifies the request and prevents a breach. To limit the impact of potential breaches, we apply segmentation policies, employ the principle of least privilege access, and use analytics to help detect and respond quickly.

Zero trust is a security strategy. It is not a product or a service but an approach to designing, adopting, and implementing defined, verified security principles.

Simplified, zero trust can be boiled down to three basic principles, as defined by Microsoft:

- **Verify explicitly**: In other words, authentication and authorization should be reverified when a user or device is determined to have changed location, when a device has changed compliance or health, or if a behavioral anomaly is detected. These should all be triggers for further identity verification.

- **Use least privilege access**: Limiting user access with just-in-time access and only the level of access or assigned role required to perform their defined tasks to protect data and productivity.

- **Assume breach**: Minimize the blast radius for breaches and employ security strategies to prevent lateral movement.

Microsoft Zero Trust is currently defined by six pillars of coverage to which each of these principles can apply:

- **Identity**: Identity is the first pillar of Zero Trust, much like a cloud defense-in-depth strategy. Access should only be granted to the people, devices, and resources authorized and required to complete a task. This aligns with the **Identity and Access** layer of the defense-in-depth model illustrated in *Figure 1.2* of *Chapter 1, Cybersecurity in the Cloud*.

- **Endpoints**: Assessing the security compliance of device endpoints is the next area to which Zero Trust applies, including **Internet of Things** (**IoT**) devices. This aligns with the **Compute** layer of the defense-in-depth model illustrated in *Figure 1.2* of *Chapter 1, Cybersecurity in the Cloud*.

- **Applications**: The entry point of access to an application should use Zero Trust oversight. Access to resources that are required for the application such as databases or data sources should also require additional verification. This aligns with the **Applications** layer of the defense-in-depth model illustrated in *Figure 1.2* of *Chapter 1, Cybersecurity in the Cloud*.

- **Network**: Zero Trust protections at the network layer for accessing resources are crucial – especially those within the corporate perimeter. This includes utilizing network and web application firewalls that provide deep packet inspection and application protection, respectively, to verify the integrity of the information that is coming into the network. This aligns with the **Network** layer of the defense-in-depth model illustrated in *Figure 1.2* of *Chapter 1, Cybersecurity in the Cloud*.

- **Infrastructure**: Applying Zero Trust principles to the infrastructure hosting your data on-premises and in the cloud is a step just beyond the network pillar. This can be done with the overall hardening of operating systems, containers, or microservices that are part of the physical or virtual infrastructure to decrease the potential attack surface. Protecting access to ports through just-in-time access and bastion hosts can also decrease the likelihood of a security breach. This aligns with the **Physical** Security and **Perimeter** Security layers of the defense-in-depth model illustrated in *Figure 1.2* of *Chapter 1, Cybersecurity in the Cloud*.

- **Data**: Protecting the integrity of your data for customers, people, and the overall business can be done through utilizing key management systems, such as Azure Key Vault, data loss prevention and data governance with Purview, and vulnerability scanning and threat detection on database platforms. This aligns with the **Data** layer of the defense-in-depth model illustrated in *Figure 1.2* of *Chapter 1, Cybersecurity in the Cloud*.

These six pillars create an end-to-end model for enforcing Zero Trust and decreasing the attack surface and the number of entry points for attackers or channels that are open to leak sensitive information. Optimizing identity and access management as the first line of defense in this strategy ensures that users are who they say they are at every access attempt, and regularly reaffirms their trustworthiness.

The next section will expand upon how the zero-trust approach should be used to design a strategy for access to resources within cloud and hybrid environments in relation to identity and access.

Designing a Strategy for Access to Cloud Resources

While determining how access to cloud resources will be handled, let us think about the evolution of identity and access architectures. Identity and access prior to cloud technologies were typically handled from application to application. Developers of on-premises applications would have their user database accessed from within the application. Every application would have a database of users and passwords. Windows AD user databases could be tied into these applications and could also be integrated into some of these applications. However, full **single sign-on** (**SSO**) capabilities were not widely available.

As cloud technologies became available and **software-as-a-service** (**SaaS**) applications became more widely used by companies, the need to manage and govern identity and access expanded beyond the on-premises architecture. This expansion decreased the amount of control that IT departments had to protect identities and access behind network technologies, such as firewalls, **virtual local area networks** (**VLANs**), and **virtual private networks** (**VPNs**).

Modern authentication requires identity and access to be managed across on-premises networks, SaaS applications (Microsoft 365, Google Workspace, formerly called G-Suite, and so on), and cloud providers such as Microsoft Azure, Amazon Web Services, and Google Cloud Platform. These environments have data that includes **personally identifiable information** (**PII**), business-critical information, and intellectual property that needs to be protected. The identity and access architecture of these environments must take a zero-trust approach to verify users and devices explicitly before authorizing access to resources. *Figure 4.1* provides a diagram that shows the relationship between the various resources to access, the access control, and privileged access for the administration of resources:

Modern authentication has the following requirements for the access controls for resources and the parameters for privileged administrative access:

- Maintain a secure level of access to all resources by explicitly validating the trust of users and devices during access requests, using all available data and telemetry.

- Ensure that security assurances are applied consistently and seamlessly across the environments, including on-premises, SaaS, and cloud providers.

- The architecture should be comprehensive while enforcing an access policy as close to the resources and access entry points as possible.

- Controls that are in place should be identity-centric. They should prioritize the use of identity and related controls when available since the physical infrastructure is no longer a responsibility when working in cloud environments.

Coordinating the various environments to access adds additional complexity to the enforcement of zero trust. Traditional on-premises identity and access directories do not have SSO integration with cloud providers and applications. In addition, these on-premises networks do not have visibility of identity threats within these expanded environments.

The architect of the identity and access requires a holistic review of the integration points in the hybrid environment. *Figure 4.1* shows the expanded view of the enterprise.

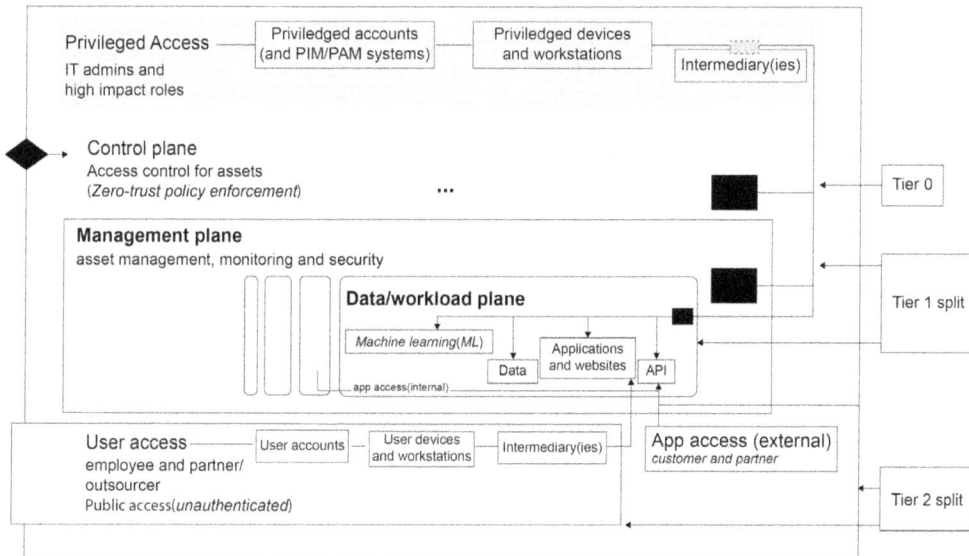

Figure 4.1: Expanded visualization of identity and access controls for secure access

The different tiers in *Figure 4.1* are described as follows:

- **Tier 0 – Control Plane**: The **control plane** is responsible for identity access controls and network access control where it is the best or only option. This tier is critical for enforcing zero-trust policies and managing privileged access to high-impact roles, ensuring that only authorized personnel have the ability to manage sensitive IT systems.

- **Tier 1 Split**: To increase clarity, business focus, and operational effectiveness, Tier 1 is split into two distinct planes:

  - **Management Plane**: This plane manages enterprise-wide IT operations, focusing on asset management, monitoring, and security of the entire IT infrastructure.

  - **Data/Workload Plane**: This is dedicated to per-workload management, focusing on protecting business-critical systems and supporting the needs of DevOps teams.

- **Tier 2 Split**: Tier 2 focuses on external app access and broader user interactions:

  - **App Access**: This handles API attack surfaces, ensuring secure external access for both customers and partners.

  - **User Access**: This covers user interactions across B2B, B2C, and public access scenarios, expanding the security management to accommodate diverse access patterns.

The various means of user and app access need to be considered when determining the verification points for a zero-trust framework. The steps to protect these access points with zero trust will be determined as follows:

- Integration and federation are created between cloud identities and on-premises identity databases for a single management plane. To be able to enforce zero trust across the enterprise, you need to have a strong modern authentication architecture.

- Microsoft Entra provides a method for enabling strong authentication and integrates endpoint security across multiple types of devices. It can also serve as the primary point for enforcing policies to verify and guarantee least-privilege access in a hybrid environment.

- Microsoft Entra offers CA to enforce policy-based decisions for access to cloud and on-premises applications based on user identity, the access environment, device health, and potential risk. These policies ensure explicit verification of identity at the point and time of access. CA policies act as a gate to verify access and identify areas for remediating potential vulnerabilities and risks. Planning and creating a strategy for CA will be discussed in the *Designing a Strategy for CA* section of this chapter.

- The visibility of user and device activity through analytics can provide operational insights into the entire identity and access infrastructure. Logging within Microsoft Entra provides reporting on authentication, authorization, and provisioning activities across users and devices. These logs can be used in Microsoft Entra for monitoring and reporting as well as integrated into SIEM solutions, such as Microsoft Sentinel, for advanced hunting of potential identity attacks on the MITRE framework.

- Identity protection and identity governance monitor and manage the access privileges of identities. Identity protection determines the potential risks to users and sign-ins. Identity governance reviews the roles of users and manages their continued access through access reviews. Microsoft's machine learning capabilities can analyze various threat signals from security solutions to improve the overall detection, protection, and response to user and device identity and access.

- Identity and access architectures that span across on-premises and cloud providers can be integrated into Microsoft Entra for a complete view and more effective management and governance. This can be accomplished with **Microsoft Defender for Identity** (**MDI**) and Microsoft Defender for Cloud Apps.

- MDI provides data on risky behavior for users that are accessing on-premises resources and applications, through either traditional or modern authentication methods. Microsoft Defender for Cloud Apps provides and discovers third-party cloud applications that are being accessed on the company network. This discovery allows you to verify and approve access to compliant applications.

- Utilizing Microsoft Entra for identity and access provides a cloud-based identity and access management service for all Microsoft cloud solutions and an integration point for on-premises and cloud-based identity providers.

- Microsoft Entra can provide SSO authentication, CA, verification with **multi-factor authentication (MFA)**, and password-less authentication. This adds new capabilities of provisioning users through automation and provides additional enterprise-enabled features for protecting and automating identity protection and governance.

The next section will discuss recommendations for an identity store and how Microsoft Entra can be used as an integration point for multiple identity providers for cloud and on-premises applications.

Recommending an Identity Store

As stated in the previous section, you need to understand the resources needed for accessing and designing an identity and access architecture with zero trust. The landscape of identity and access expands beyond the company member users. You need to understand the users who will be guests in your tenant, the companies that partner with your company for collaboration, and the customers who will be accessing your company resources, such as an e-commerce website or registration pages for events.

These scenarios pose potential risks to the organization as the identity and access environment expands beyond your own tenant. The foundational security policies and techniques used to protect member users should be maintained by anyone accessing the organization's resources. Users' identities, business partners, and customers should be protected through the security capabilities available within Microsoft Entra.

Overall, users want flexibility when accessing resources. Microsoft Entra provides these capabilities through foundational security features, such as MFA. Expanded features to provide SSO for members, guests, and customers can be provided and allow access to applications that are published and verified through Microsoft Entra. The following sections will discuss the diverse options for providing this flexibility in Microsoft Entra.

Microsoft Entra Tenant Synchronization with SCIM

For large companies that partner and need to federate with your tenant, the **System for Cross-Domain Identity Management (SCIM)** protocol can be used. SCIM is an open-source protocol that can be configured with a user management API to synchronize and automatically provision users and groups to an application in Microsoft Entra. *Figure 4.2* provides a diagram of how this works across multiple identity providers:

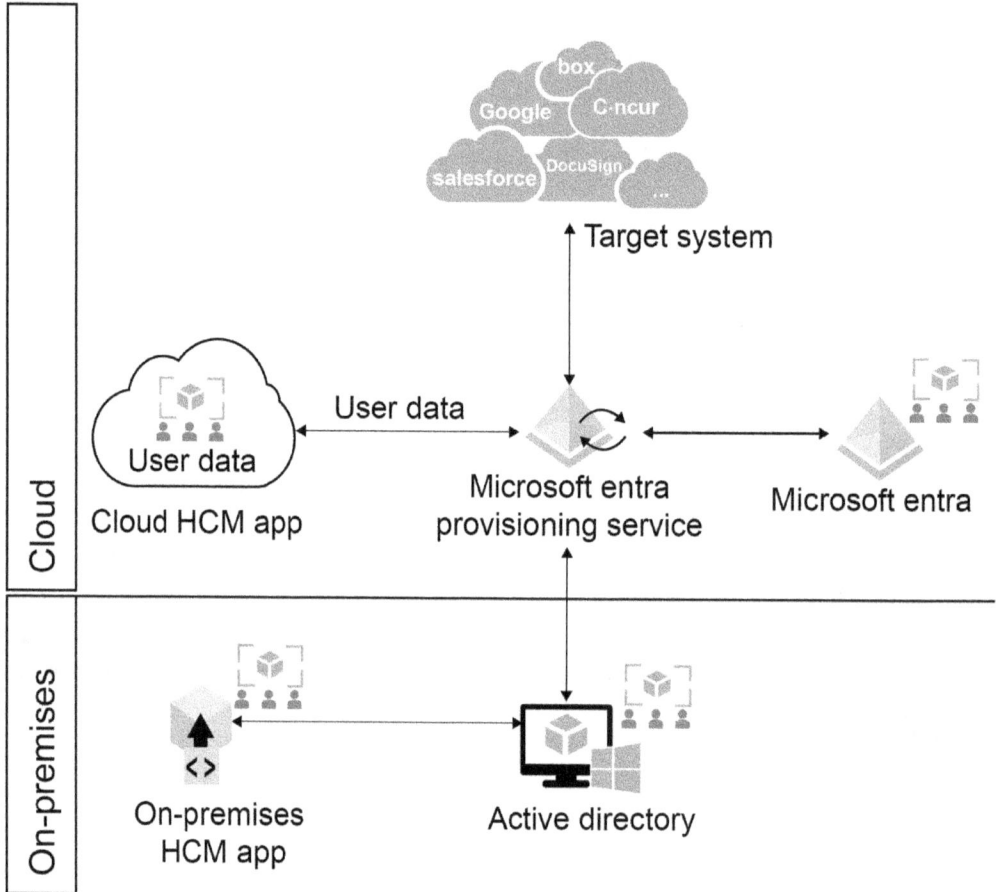

Figure 4.2: SCIM Microsoft Entra provisioning diagram

> **Note**
>
> For more information on how to use SCIM in Microsoft Entra, please go to the following link: https://learn.microsoft.com/entra/architecture/sync-scim.

SCIM synchronization with Microsoft Entra is a good option for integrating companies that need cross-platform access to resources in a merger and acquisition scenario. For other partner relationships between two companies, **business-to-business** (**B2B**) access can be used.

External Identities

B2B

B2B guests are best described as a partnership relationship between users within two separate companies that need to collaborate on a project. These B2B relationships may be created through business mergers and acquisitions, project needs, or support relationships.

Within these B2B relationships, an external company may bring its own Microsoft Entra and Microsoft 365 licenses for collaboration. Otherwise, these licenses can be assigned if external users do not come from a Microsoft Entra tenant with Microsoft 365 licensing. External company collaboration settings provide these users with an SSO experience for both business tenants. If an external user does not use these licenses, then the username and password can be used from another identity, and that user can be assigned licenses from within the invited tenant.

There are several variations of B2B functionality available, depending on your specific organizational requirements.

B2B Collaboration

This allows you to collaborate with external users by using their chosen identity and **identity provider** (**IdP**) rather than your preferred IdP and results in those users being represented in your tenant/ directory as guest users.

The onboarding flow will typically be one of the following:

- Invite the external users to access specific applications within your tenant.
- Self-service signup – the users can sign up for access to an application in your tenant using their preferred identity.
- Use Microsoft Entra ID **entitlement management** (**EM**), which allows you to manage external users with external identities in a more scalable fashion using automation, access reviews, automatic expiration of access, and more.

You can also leverage **cross-tenant access settings** (**CTAS**) to control which external tenants that also use Microsoft Entra/Azure can access your tenant and which tenants your own users can access.

If those other users are in non-Microsoft tenants, you can use the **External Collaboration Settings**.

> **Note**
>
> You can and should read in more detail about the capabilities and configuration options at the following links:
>
> `https://learn.microsoft.com/en-us/azure/active-directory/external-identities/what-is-b2b`
>
> `https://learn.microsoft.com/en-us/azure/active-directory/governance/entitlement-management-overview`
>
> `https://learn.microsoft.com/en-us/azure/active-directory/external-identities/cross-tenant-access-overview`
>
> `https://learn.microsoft.com/en-us/azure/active-directory/external-identities/external-collaboration-settings-configure`

B2B Direct Connect

B2B Direct Connect currently only works with Microsoft Teams shared channels and enables a more seamless experience for teams' collaboration.

This does require some preparation in advance by the admins of both tenants to enable a two-way trust relationship. Once this has been configured, the experience for users is seamless. The user connecting to a channel in another tenant authenticates as normal in their own tenant and, in turn, receives a token from the remote tenant granting them access.

While the external users do not show in your home tenant directory, you can search for users in that remote tenant.

For example, if you enabled this with Microsoft (which is a common use case), once it is set up, you can search for and chat with any user in that remote tenant (in this case, a Microsoft Entra ID tenant).

As with B2B collaboration, CTAS provides much of the control.

> **Note**
>
> You should also refer to the following link for a more detailed and current list of capabilities and limitations: `https://learn.microsoft.com/en-us/microsoftteams/deploy-chat-teams-channels-microsoft-teams-landing-page`.

B2B provides the flexibility of access to resources for business collaborations. For customers who would like flexibility in completing forms or shopping on e-commerce sites, Microsoft Entra can be used with registered applications through the configuration of **business-to-consumer (B2C)** access.

B2C

The B2C relationship is an external account used for customers accessing applications and resources within the tenant. An example of this would be using LinkedIn, Facebook, or Google credentials to log in and access an account on a shopping site. This provides convenience to customers by not requiring them to create another username and password. These IdP relationships enable B2C authentication relationships. This is more commonly referred to as a **customer identity and access management (CIAM)** solution.

Multi-Tenant Organization Feature

While not yet in the exam curriculum at the time of publishing, due to this feature still being in preview and not generally available, it is advantageous to be aware of the **Multi-Tenant Organization (MTO)** feature.

This feature is similar in concept to Active Directory Trusts and allows you to synchronize multiple tenants that all belong to the same business entity. For example, you work for a large multi-national organization with many different geo-located business entities, each with its own infrastructure and identity providers. To enable a more seamless collaboration experience across a business such as this, MTO may be a better solution than B2B/B2C capabilities alone to enable a more seamless collaboration experience.

> **Note**
> You can learn more about this at this link: `https://learn.microsoft.com/en-us/ entra/identity/multi-tenant-organizations/`.

Choosing an Appropriate External Identity and Collaboration Solution

When choosing between the plethora of external identity and collaboration solutions that are available in Entra ID, you will find it useful to study and learn the differences between each.

> **Note**
> There is a useful feature comparison table that you can access at this link: `https://learn. microsoft.com/en-us/entra/identity/multi-tenant-organizations/ overview#compare-multitenant-capabilities`.

When architecting identity and access with Zero Trust, the more that you can integrate the security features of Microsoft Entra into the methods for accessing resources within your tenant, the better. The methods described in this section provide these capabilities to enforce Zero Trust features, such as MFA and CA.

In the next section, you will learn more about choosing a strategy for authentication.

Recommending an Authentication and Authorization Strategy

In current organizational infrastructures, authentication and authorization to resources are not limited to cloud-only users. Many companies have applications that are still in on-premises data centers that users require access to. This provides additional challenges to enforcing the modern authentication techniques for zero trust. When using Microsoft Entra for authentication and authorization to cloud resources, you should also determine the proper techniques for users to access on-premises resources. Microsoft Entra Connect provides this capability for SSO to cloud and on-premises resources, but you need to determine the best method for your company to synchronize and manage these hybrid users.

Hybrid Identity Infrastructure

The term *hybrid identity* is meant to signify that the company has users that use on-premises resources, and users that use cloud-native resources. Within this hybrid identity infrastructure, there is going to be an on-premises Windows AD domain controller that is used to manage the on-premises users, and Microsoft Entra, which manages the cloud-native users, both members and guests. This infrastructure coincides with companies that have a *hybrid cloud* approach. Many companies have Windows AD domain controllers in place today. Microsoft Entra Connect provides a hybrid infrastructure connection to Microsoft Entra.

Microsoft Entra Connect is a software solution that is installed within the on-premises infrastructure and configured to synchronize users and groups to Microsoft Entra. Microsoft Entra Connect simplifies the management of these users and groups by providing ways that an identity and access administrator can manage users in one interface and have the changes updated in near-real time.

Since there are structural differences between how **Active Directory Domain Services (AD DS)** and Microsoft Entra are built, Microsoft Entra Connect provides the means to create a consistent user and administrator experience for identity and access management.

> **Note**
>
> There are some prerequisites and aspects that are out of scope within Microsoft Entra Connect that you should understand. The installation prerequisites can be found at this link: `https://learn.microsoft.com/entra/identity/hybrid/connect/how-to-connect-install-prerequisites`.

When planning to implement Microsoft Entra Connect, the following information should be understood and planned accordingly:

- Microsoft Entra Connect synchronizes users and groups, not devices or applications. There are ways to co-manage devices and support on-premises application access within Microsoft Entra. Both will be covered in later chapters.

- Microsoft Entra Connect synchronizes a single AD DS forest per Microsoft Entra tenant. If there are multiple forests, then multiple tenants will be required in Microsoft Entra.

There are additional considerations in planning for AD DS and Microsoft Entra synchronization with Microsoft Entra Connect, but these will be covered within the use cases for each synchronization type.

There are three options when configuring Microsoft Entra Connect for synchronization:

- **Password hash synchronization (PHS)**

- **Pass-through synchronization (PTS)**

- Federation with **Active Directory Federated Services (AD FS)** synchronization

Each of these options has unique uses. The following subsections will detail these uses and options to consider when choosing one to use within your company.

PHS

PHS is the easiest to configure and is the default option within Microsoft Entra ID Connect express setup. PHS maintains both the on-premises and cloud identities of users. This takes place by providing on-premises user identities to Microsoft Entra along with an encrypted hash of their passwords. This allows users to sign into on-premises and cloud applications with the same authentication credentials.

PHS is a good option when a company has a single on-premises domain and is moving quickly to a cloud-native infrastructure. PHS is not for companies with complex authentication and password requirements within an on-premises AD.

As previously stated, PHS maintains authentication credentials on-premises and in Microsoft Entra. Therefore, PHS can have users authenticate to cloud applications through Microsoft Entra, while passing authentication responsibilities to on-premises applications to on-premises AD. The benefit here is that if the connection fails in Microsoft Entra Connect between the on-premises AD DS and Microsoft Entra, users are still able to authenticate to their cloud applications and remain partially productive.

Figure 4.3 shows how this workflow is handled:

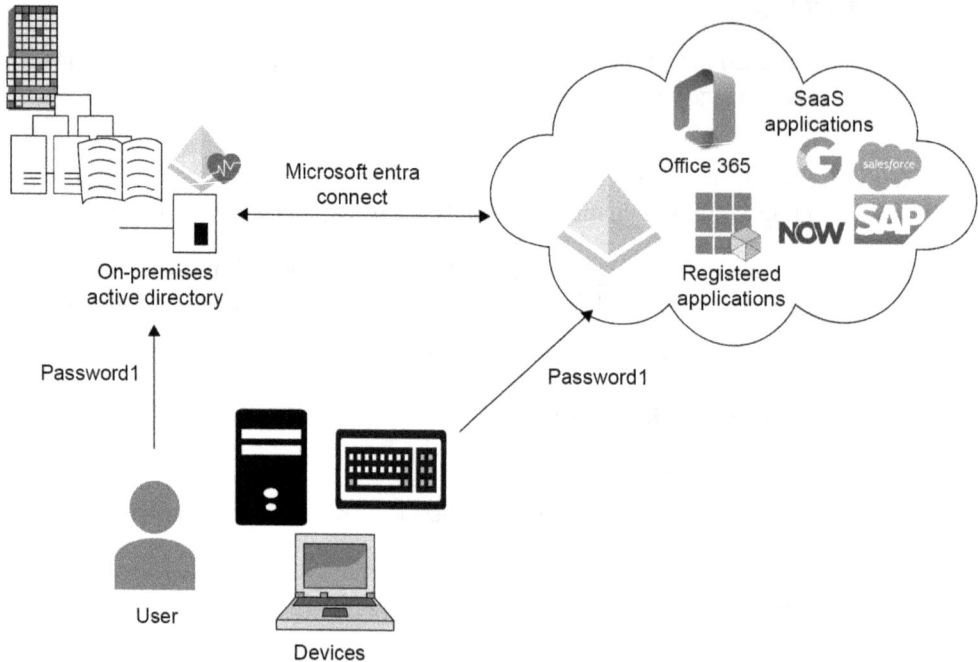

Figure 4.3: Overview of PHS, showing the flow from user devices to
on-premises Active Directory and Microsoft Entra Connect

Additional information can be found at this link: `https://learn.microsoft.com/entra/identity/hybrid/connect/whatis-phs`.

This configuration for Microsoft Entra Connect is the least complex of the three options and should be preferred for a cloud-native authentication approach. The next sections will provide information about PTS and AD FS synchronization.

PTS

The next hybrid identity synchronization option is PTS. Unlike PHS, which allows user identities to be authenticated in either the on-premises AD or Microsoft Entra, PTS requires all users to authenticate to the on-premises AD.

In this configuration, if the Microsoft Entra Connect connection between Microsoft Entra and on-premises AD were to become disconnected, no users would be able to authenticate to on-premises or cloud resources. Therefore, it is important to actively monitor this connection and build resiliency in the architecture. *Figure 4.4* shows a diagram of how PTS functions and how you can build resiliency with redundancy in **pass-through agents (PTAs)** and a backup domain controller:

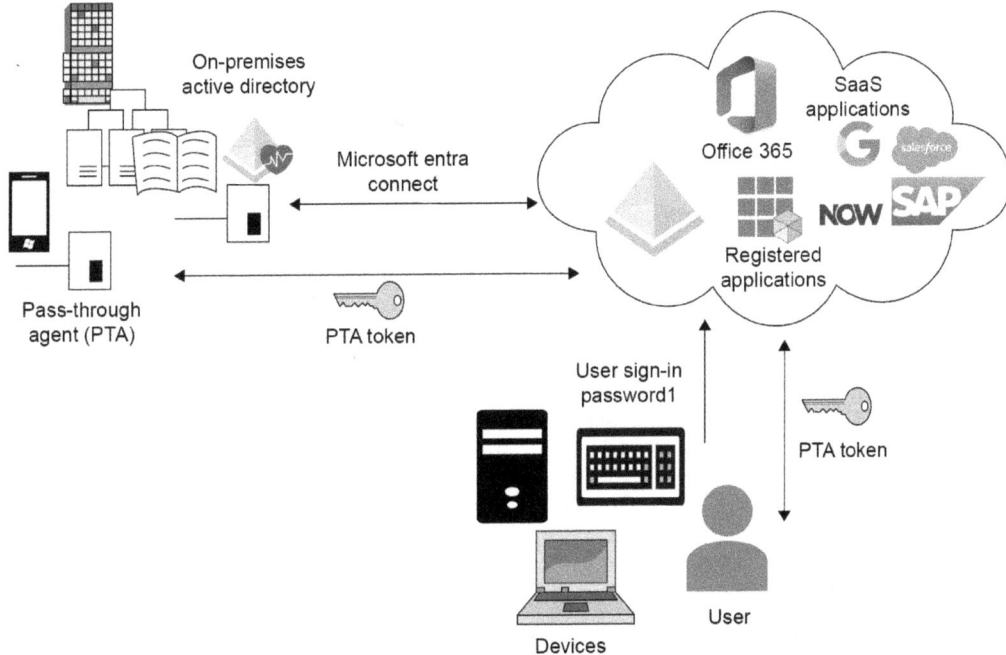

Figure 4.4: An overview of PTS, with the key difference from PHS being the dependency
on communication between on-premises Active Directory and Microsoft Entra

There are some reasons to utilize PTS. An organization may require authentication parameters and limits that only allow users to access resources during certain times. Such rules can only be configured currently on an AD domain controller. With PTS, you can utilize modern authentication features with Microsoft Entra, such as MFA and **self-service password reset** (**SSPR**). However, you must enable the password writeback feature within Microsoft Entra Connect to use SSPR. Password writeback will allow the Microsoft Entra password to be written to the on-premises AD. If this is not enabled, the AD password will always take precedence.

To have a resilient architecture, it is recommended that at least two of these PTAs are installed on member servers in the on-premises infrastructure. Microsoft's recommendation is that up to four PTAs should be deployed. The next section will discuss the third and final synchronization type within hybrid identity infrastructures, AD FS.

Federation with AD FS Synchronization

Federation with AD FS synchronization is the most complex of the three Microsoft Entra Connect synchronization types. AD FS requires additional infrastructure in place to support the authentication process. In comparison, PHS and pass-through synchronization can be installed directly on the on-premises domain controller in many cases. *Figure 4.5* provides an overview that shows the complexity of the infrastructure and necessary components:

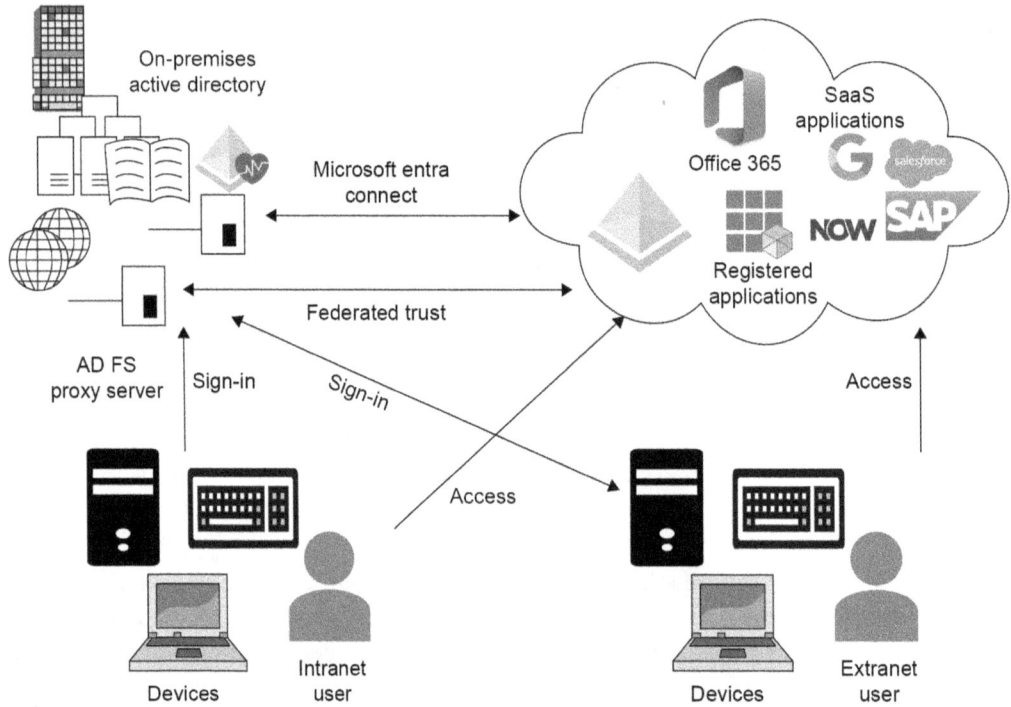

Figure 4.5: An overview of AD FS synchronization flows

AD FS synchronization is utilized in complex AD infrastructures where there are multiple domains, and third-party MFA solutions or smart cards are utilized.

> **Note**
>
> For additional information on the configuration of AD FS synchronization, you can read more here: `https://learn.microsoft.com/entra/identity/hybrid/connect/how-to-connect-fed-management`.

PHS and PTS are the more widely discussed of the three Microsoft Entra Connect options. Federation with AD FS is necessary when utilizing third-party solutions that are not native to Microsoft, such as third-party MFA.

The next section will discuss additional methods for secure authorization that should be considered when designing identity and access security with zero trust.

Secure Authorization Methods

Once you have determined the method that you will be using for authentication within your hybrid infrastructure, you should consider how to authorize users to access resources. The synchronization methods for authentication allow you to utilize Microsoft Entra security methods for authorization. These methods include the following:

- **Security group membership**: Planning for users to be assigned roles and authorized access based on group membership decreases the management overhead for identity and access administrators

- **Role-based access control**: Authorization for access to resources should be based on the role of the user and the level of access should be specific to the tasks that they need to perform

Properly assigning these authorization methods allows a more secure environment for managing and monitoring access to resources.

The next section will discuss how to use CA for the dynamic enforcement of zero trust.

Designing a Strategy for CA

CA policies enforce additional verification actions based on a signal that a user or device may be potentially compromised. The foundation of CA policies is the zero-trust methodology. Microsoft Entra CA analyzes signals such as user, device, and location to automate decisions and enforce organizational access policies for the resource. CA policies allow you to prompt users for MFA when needed for security and stay out of the user's way when not needed.

As you will see in *Figure 4.6*, the policies that are determined for the company are what then enforce these CA requirements from signal to decision to enforcement:

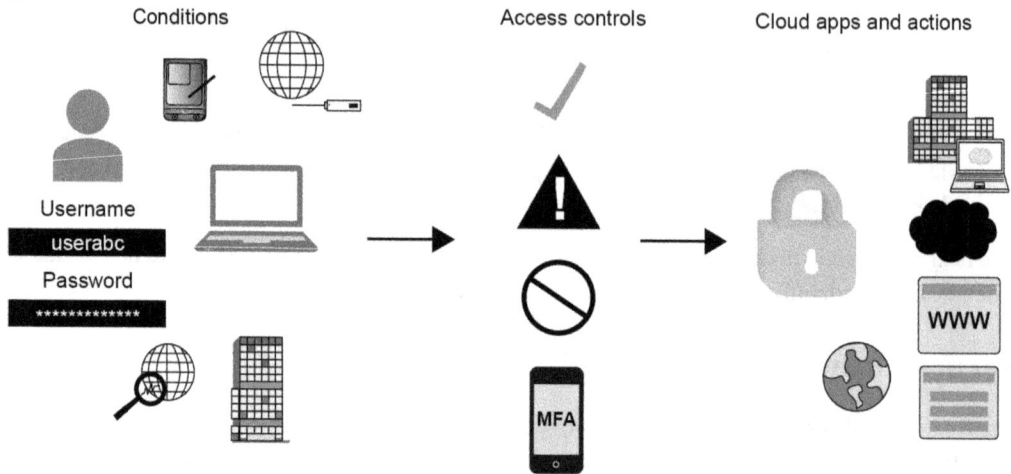

Figure 4.6: A CA workflow showing the flow of evaluation of user access, from left to right

The planning and creation of CA policies should be a foundation of access policy enforcement in zero trust. In addition, you should have a set of active and fallback policies to start your deployment. You need to have a proper plan and understand how CA policies would potentially affect the user experience. There is a balance that a company should attempt to maintain between the enforcement of policies to secure and protect data. This includes the ability for a user to have access to the applications and data that they need to be effective at their required tasks.

Before you can create a CA policy, you will need to meet the prerequisites. There are a couple of areas that you need to address to implement this solution, that is, licensing and security defaults. For licensing, CA policy features are available with a Microsoft Entra Premium P1-level license. This level of Microsoft Entra licensing includes Microsoft 365 Business Premium, Office 365 E3/A3, Microsoft 365 E3/A3, Office 365 E5/A5, and Microsoft 365 E5/A5. These licenses must be assigned to the users for whom we are trying to enforce CA policies.

> **Note**
>
> The full list of licensing requirements can be found at this link: `https://learn.microsoft.com/entra/identity/conditional-access/overview`.

In addition to the proper licenses, we must turn off the Microsoft Entra security defaults. Security defaults are turned on when we create our Microsoft Entra tenant. This provides a baseline level of protection. Examples include when users enroll in MFA, enforce MFA for administrators, and block the use of legacy authentication for apps. Security defaults will be discussed further in *Chapter 7*. To be able to implement CA, navigate back to **Security Defaults** and turn them off. When you do this, there will be a list of reasons that appear, and you will see the selection for using CA policies.

Once you have the proper licensing assigned and security defaults turned off, you can begin planning for CA policies.

> **Note**
>
> Some commonly used CA policies can be found on the Microsoft Docs website at this link: `https://learn.microsoft.com/entra/identity/conditional-access/plan-conditional-access`.

The key to planning for CA is to understand the following:

- The groups of users that access company applications and data
- The devices that they are using to access company applications and data
- The locations where they may be accessing company applications and data
- The applications that are being used to access the company data

It is essential to understand how **Conditional Access policies** (**CAPs**) are evaluated to ensure they operate in the manner you expect and that you do not have unintended gaps in your protection.

CAPs are not evaluated in isolation and are also not evaluated like, for example, a firewall ruleset, where evaluation terminates after the first **Allow** action.

Instead, it can be thought of in a similar fashion to **Group Policy Objects** (**GPOs**), which you may be familiar with from on-premises Active Directory.

If you are not familiar with Active Directory, an analogy is the configuration of **access control lists** (**ACLs**). ACLs are used to define permissions for users and groups on files and folders. The system evaluates all the ACL entries to determine the effective permissions for a user.

Multiple CAPs can apply to your session simultaneously:

1. **Data collection**: CA collects session details such as your location and device.
2. **Policy matching**: It reviews which policies match your session conditions.
3. **Block policy check**: If any block policy condition is met, access is denied immediately.

4. **Requirement evaluation**: If no block conditions are met, the remaining policies are combined into a single set of requirements.

5. **Additional requirements**: If these policies require additional actions (e.g., MFA and compliant device), these are checked. If any requirement is not met, access is denied.

6. **Session controls**: If all requirements are met, session controls (e.g., sign-in frequency) are applied. The most restrictive control is used if multiple policies define the same control.

For example, if one policy sets a sign-in frequency of 12 hours and another policy sets it at 8 hours, the 8-hour limit will apply.

With the many overlapping conditions in a single policy and the ability for multiple policies to be aggregated and applied at once to a single session, it is possible that they may not function as intended.

Therefore, when a new CA policy is created, you should test it in report-only mode using the **What-If** feature to verify that the policy is working correctly against users, devices, and applications.

What-If testing will also show other policies that may overlap and cause a conflict with user access.

Microsoft Entra Identity Protection

Microsoft Entra Identity Protection is an additional capability that requires the Microsoft Entra ID P2 license (per user) or Microsoft 365 E5 license (per tenant) and provides additional capabilities to protect your riskiest users.

It analyzes signals from all Microsoft tenants and services, including Entra, Microsoft 365, Xbox, and Microsoft accounts, to provide insights into user and sign-in behavior. This includes detecting the use of anonymous IP addresses, impossible travel (e.g., signing in from London, United Kingdom, and then another country such as the United States within an impossible timeframe), malware-linked IP addresses, appearances on password breach databases, password spray attempts, and more.

Based on those insights, Microsoft will assign risk scores to users and their sign-ins. These scores will be accessible through the Entra and Azure portals, sign-in logs, the Graph API, and CA.

The CA integration allows an administrator to also evaluate user risk and sign-in risk in CAPs and take required enforcement actions if one or both scores are above a particular value (low, medium, or high). For instance, this could require a user to change their password, re-authenticate using MFA, adjust their sign-in frequency, or even block their access.

Designing a Strategy for CAE

CAE is a concept that recognizes the need to be able to, as quickly as possible, revoke or amend access based on changes in the posture of a user, their identity, their session, their device, and other attributes that contribute to their overall security posture.

Microsoft Entra uses the OpenID decentralized authentication protocol within Entra, which enables many of the use cases discussed in this chapter, such as seamless SSO, across multiple applications, using a single credential.

When you sign in to Entra, a refresh token is created that authorizes you to access authorized resources for the life cycle of that token.

Shorter token expiration periods can lead to user frustration through excessive authentication attempts; however, the longer the token's validity, the greater the security risk presented.

> **Note**
>
> The way that Microsoft meets this challenge is through the adoption of CAE, which is itself built on the OpenID Continuous Access Evaluation Profile standard, defined in detail at this link: `https://openid.net/specs/openid-caep-specification-1_0.html`.

In practice, this allows Microsoft services using OpenID and refresh tokens to frequently ask Microsoft Entra (the IdP) whether your user posture has changed and whether to continue to allow or block access. Consider the scenario where a credential might be found to be compromised/stolen. You disable the account but the **primary refresh token** (**PRT**) in Entra is valid for 14 days. Without CAE, the malicious actor who has access to the stolen credential can continue to use it within your infrastructure for up to 14 days (about 2 weeks).

With CAE, Entra and related Microsoft services such as Teams, Exchange Online, and SharePoint subscribe to critical events such as disabling/deleting user accounts, password changes, administrative session revocation, and so on. When these occur, the subscribing services are alerted and will revoke access in near-real time.

The services mentioned also subscribe to changes in CAPs in near-real time.

CAE is enabled by default for new tenants; existing tenants may need to first enable it and undergo a migration process.

> **Note**
>
> You can find more details at this link: `https://learn.microsoft.com/en-us/entra/identity/conditional-access/concept-continuous-access-evaluation`.

Protected Actions

An additional capability to help secure privileged access is the protected actions capability.

> **Note**
>
> This is documented in detail at this link: `https://learn.microsoft.com/en-us/entra/identity/role-based-access-control/protected-actions-overview`.

This provides more granular targeting of CAPs beyond just users, groups, applications, and locations; it enables you to set a policy for the following categories of protected actions:

- CAP management
- Cross-tenant access settings management
- Custom rules that define network locations
- Protected action management

As these controls are key security controls, it is wise to add extra layers of protection to them, such as enforcing re-authentication with MFA, requiring stronger authentication methods in MFA, and so on. This aligns with the defense-in-depth strategy covered in *Chapter 1*, but also with zero-trust principles.

As we continue to design and plan access to resources, the assignment of roles and delegation of roles to resources in a zero-trust architecture is needed. The next section will discuss how to design a strategy for role assignments and delegation to access resources.

Designing a Strategy for Role Assignment and Delegation

Managing role assignments and how users are delegated access to resources can be a source of concern for security and governance. To alleviate the issue of users requesting access and being assigned a role on a one-to-one basis, planning a strategy for role assignments by groups of users creates a more manageable environment.

Security groups can be created for departments or project groups that include dynamic assignments. These groups can be assigned the roles defined by management and supervisor stakeholders for members of the groups. When a user is then manually or dynamically assigned to that group, they inherit that role. When users are removed from the group, the role is no longer available to them. This allows you to better manage and govern the access levels and roles that users in your company have to resources.

> **Note**
>
> The differences between Microsoft 365 groups and security groups can be found at this link: `https://learn.microsoft.com/microsoft-365/community/all-about-groups`.

This becomes especially important when you are planning for the privileged roles of administrators in your tenant. The next section will discuss the planning of this strategy for these elevated user privileges.

Designing a Security Strategy for Privileged Role Access

Designing a secure strategy for privileged roles will allow you to protect and defend identity and access by utilizing the concept of zero trust and the principle of least privilege to assign authorization for administrator accounts. You should have a clear strategy with defined job tasks for every administrator user account to plan for the proper assignment of these roles. This strategy should include meeting with stakeholders and discussing the roles that each department member requires to complete their job tasks. In addition, you should be monitoring the activity of these accounts and verifying the continued requirement for users to have these privileged access roles.

To enforce the concept of zero trust, you can assign CAPs to these accounts. To address and protect privileged assignments, Microsoft Entra provides **Privileged Identity Management** (**PIM**) within the **Identity Governance** solutions.

Microsoft Entra ID PIM

PIM provides just-in-time privileged access to users. Since users are only provided active administrator roles for a short window of time, this reduces the attack surface and potential for these user accounts to cause exposure to privileged access in an attack. PIM provides an approval and justification process for activating privileged role assignments, which includes notifications when a role is activated and an audit trail of these activations.

PIM requires a Microsoft Entra Premium P2 license. To assign PIM to member accounts, each user must have this license. However, for guest users that require privileged access with PIM, five guests can be assigned PIM roles for each Microsoft Entra Premium P2 license that you have in your tenant.

When planning for Microsoft Entra PIM, you should consider the following options:

- Roles should be assigned utilizing the principles of least privilege. The assignment should use a role that provides the authorized level of access necessary to perform the task and no more.

- PIM should be used for just-in-time access to roles and should only be active for a specified and limited amount of time. Administrator roles should not be permanently assigned; they should be eligible by request.

- Administrator accounts should have MFA enabled and enforced without exceptions. These accounts should also be cloud-native accounts and not accounts that are synchronized with Windows AD.

- Recurring access reviews should be used to verify that access to the assigned roles is still necessary for the users and groups.

- Global administrators within the company should be limited to five users. This protects the attack surface in the case of a compromised account and protects against users who may leave the company with elevated privileges.

To protect against a potential lockout and ensure access is still available in a potential emergency, you should configure at least two emergency access or "break-glass" accounts. These accounts are accounts of high privilege with access at the level of a global administrator. These accounts are not protected with MFA, so they can gain access quickly to resources when other administrator accounts cannot gain access. These accounts should be limited to this scenario, and the credentials should be locked away until they are needed.

Break-glass accounts are member accounts tied directly to the Microsoft Entra tenant. Therefore, they can be utilized in situations where federated IdPs are being used for authentication and there is an outage to that IdP. Other use cases would be that the global administrator has lost access to their MFA device to verify their identity, a global administrator has left the company and it is necessary to delete that account, and a storm has taken down cellular services and you cannot verify the identity with MFA.

> **Note**
>
> Additional information on emergency access or break-glass accounts can be found at this link: `https://learn.microsoft.com/entra/identity/role-based-access-control/security-emergency-access`.

The next section will provide additional ways to manage and govern the activities of privileged users and their access to resources.

Designing a Security Strategy for Privileged Activities

In terms of the access life cycle, you should consider the access life cycle of your member users, your guest users, and especially your privileged users. These should be handled differently as the life cycle of member users is based on their employment within the company and the access required for the department or team they belong to. Guest users are provided access based on a partnership and an external collaboration trust relationship.

Privileged Access Reviews

Privileged user access should be regularly reviewed in a comparable manner. Since these are elevated access assignments, these reviews should be done consistently as identified by the company. Unused and unnecessarily privileged assignments should be removed. Automated removal should also be configured for users who are no longer with the company or have changed departments within the company. In the next section, you will learn how entitlement management can be used to govern access to resources for internal and external users.

Entitlement Management (aka Permission Management)

Entitlement management provides this governance through the creation of catalogs and access packages that you can build for these groups of users. Entitlement management is found under **Identity Governance** within Microsoft Entra ID. In entitlement management, the catalogs define the groups and teams, applications, and SharePoint sites within Identity Governance. Creating a catalog does not establish access to these catalogs. You must go through the creation of access packages to approve and allow access to these catalogs.

Before creating catalogs and access packages, you should plan and determine how these are going to be used within your company. Entitlement management can be a helpful tool for companies that have projects that utilize internal and external users, departments that utilize different and specialized resources that other departments do not require access to, and branch and global offices that have their own users, groups, and partners.

Overseeing identity governance, it is important to work with stakeholders to plan these catalogs and access packages and to determine how often they will be reviewed for continued use and access. Proper planning with these stakeholders will allow them to quickly provide users with access to the resources that are required for a given project or department once they are onboarded.

Like privileged access reviews, entitlement management can use access reviews to verify continued membership to access packages.

The next section will discuss how to bring together the use of privileged access management, access reviews, and role permissions to administer identity and access across your tenant.

Cloud Tenant Administration

The previous sections discussed many of the strategies that can be used to enforce and verify user access through zero trust. This is a multi-staged strategy across your tenant to plan and implement. As a cybersecurity architect, you should work with the various stakeholders to meticulously plan and implement these strategies in stages.

These stages can be broken down and prioritized as follows:

- *Stage 1*: Critical items that are recommended to be done right away. This stage should be executed in the first 24–48 hours of creating your cloud tenant. Steps within this stage include the identification of privileged roles that should use PIM and the creation of break-glass accounts.

- *Stage 2*: Mitigate the most frequently used attack techniques. In this stage, you should spend 2–4 weeks addressing the common attacks, such as brute-force and phishing attacks. You should determine a proper hybrid identity synchronization strategy and set up identity protection to recognize potential identity attacks. The incident response process should begin to be developed with the assignment of owners.

- *Stage 3*: Build visibility and full control of administrator activity. The next 1–3 months should begin to see the full control of PIM for administrators, access reviews, and utilizing standards, such as those of the **National Institute of Standards and Technology** (**NIST**), for recommendations on incident response procedures.

- *Stage 4*: Continue building defenses to further harden your security platform. This stage is ongoing for reviewing and improving security controls and validating response plans. In this stage, you should determine ways to manage devices with endpoint management, and you should have all company-owned devices compliant and Microsoft Entra joined.

> **Note**
> Additional information on these stages can be found at this link: `https://learn.microsoft.com/entra/identity/role-based-access-control/security-planning`.

The next section will provide a case study that summarizes the zero-trust architecture discussed in *Chapters 2, 3,* and *4.*

Case study – Designing a Zero-Trust Architecture

Apply what you learned in this chapter by completing the case study on the accompanying online platform. In this case study, you will be given a company scenario and asked to complete several tasks to meet the requirements of the zero-trust architecture.

To access the case study, visit the following link or scan the QR code.

Link to the case study: `https://packt.link/SC100-E2-CaseStudy_Chapter4`

QR code:

Figure 4.7: QR code to access case study for Chapter 4

Summary

In this chapter, you explored the design and strategy for creating a zero-trust architecture for identity and access. This included an overview of zero-trust for identity and access management and how to design a strategy for access to cloud resources. You then learned ways to recommend an identity store for hybrid and guest access and recommend an authentication and authorization strategy. You also explored the capabilities of Microsoft Entra ID protection and CAE in providing near-real-time signals about user identities and their sign-in behavior that enable risk-based decisions to be made during authentication and authorization.

Finally, you learned about the various strategies for designing CAPs, determining role assignments and delegation, handling privileged role access, and reviewing and governing privileged activities. We then wrapped up the chapter with a case study to provide design and architecture suggestions for zero trust for users, devices, and networks.

In the next chapter, you will learn how to design a strategy for regulatory compliance.

Exam Readiness Drill – Chapter Review Section

Apart from mastering key concepts, strong test-taking skills under time pressure are essential for acing your certification exam. That's why developing these abilities early in your learning journey is critical.

Exam readiness drills, using the free online practice resources provided with this book, help you progressively improve your time management and test-taking skills while reinforcing the key concepts you've learned.

How to Get Started

1. Open the link or scan the QR code at the bottom of this page.

2. If you have unlocked the practice resources already, log in to your registered account. If you haven't, follow the instructions in *Chapter 11* and come back to this page.

3. Once you have logged in, click the **START** button to start a quiz.

We recommend attempting a quiz multiple times till you're able to answer most of the questions correctly and well within the time limit.

You can use the following practice template to help you plan your attempts:

Working On Accuracy		
Attempt	Target	Time Limit
Attempt 1	40% or more	Till the timer runs out
Attempt 2	60% or more	Till the timer runs out
Attempt 3	75% or more	Till the timer runs out
Working On Timing		
Attempt 4	75% or more	1 minute before time limit
Attempt 5	75% or more	2 minutes before time limit
Attempt 6	75% or more	3 minutes before time limit

The above drill is just an example. Design your drills based on your own goals and make the most of the online quizzes accompanying this book.

First time accessing the online resources? 🔒

You'll need to unlock them through a one-time process. **Head to** *Chapter 11* **for instructions.**

Open Quiz

https://packt.link/SC100_CH04

Or scan this QR code →

5

Design a Regulatory Compliance Strategy

The previous chapter discussed how to design an identity security strategy for cloud-native, hybrid, and multi-cloud identity and access management infrastructures. This chapter will discuss how to design security and governance strategies based on regulatory compliance requirements within your company. This includes how to utilize Microsoft Defender for Cloud and Azure Policy to evaluate and govern your company resources. A **regulatory compliance strategy** is a crucial aspect of designing security and governance strategies based on regulatory compliance requirements within your company. This skill involves understanding and implementing various compliance requirements, such as company standards, government standards, regulatory standards, and industry standards. It also includes using tools such as Azure Policy and Microsoft Defender for Cloud to monitor and manage compliance. Focusing on regulatory compliance is essential because it ensures that your company adheres to legal and industry standards, which can prevent financial penalties, business limitations, or even criminal charges. It also helps in maintaining the security and privacy of data, which is crucial for building trust with customers and stakeholders. From an exam perspective, understanding the regulatory compliance strategy will help you answer questions related to designing security and governance strategies, interpreting compliance requirements, and using tools such as Microsoft Defender for Cloud and Azure Policy. It will also prepare you for case studies and practical scenarios where you need to apply these concepts to real-world situations.

This chapter addresses the **Design a regulatory compliance strategy** domain and focuses on the following:

- Interpreting compliance requirements and translating them into specific technical capabilities
- Evaluating compliance by using Microsoft Defender for Cloud
- Interpreting compliance scores and recommending actions to resolve issues or improve security

- Designing an implementation of Azure Policy
- Designing for data residency requirements
- Translating privacy requirements into requirements for security solutions
- Case study – designing for regulatory compliance
- Knowledge base questions

Interpreting Compliance Requirements and Translating Them into Specific Technical Capabilities

When moving to cloud technologies for the access and use of company resources, there is a clear shift in a company's ability to monitor and manage compliance. When resources are within the walls of a company-owned data center, security controls, processes, and policies can be put in place to govern compliance with company, government, industry, and regulatory requirements. When discussing compliance requirements, the conversation should be based on the processes, procedures, policies, and controls that are in place for adhering to these standards. The focus of this chapter will be on the policies and controls within the technical landscape to monitor and manage compliance.

Before discussing how to interpret these compliance requirements into technical capabilities, let's define what each of these compliance types is and how they can affect how you design your infrastructure:

- **Company standards** are requirements that are set forth by the stakeholders and executives of the company. These standards are operational and could be around budget and cost, geographic use, patching and updates, or other technical requirements. Unlike the other standards, failure to meet these requirements would not carry any penalties from a government or standards enforcement group. However, not adhering could have negative effects on your company's employment status.

- **Government standards** encompass requirements set by local, state, or national governing bodies. These standards focus on proper data handling, privacy protection (especially **personally identifiable information (PII)**), and data sovereignty. Some countries strictly mandate that data processed within their borders must not leave the country. Non-compliance with these standards can lead to financial penalties, business limitations, or even criminal charges. Notable examples include the **General Data Protection Regulation (GDPR)** in the **European Union (EU)** and the **United Kingdom (UK)**, as well as the **National Institute of Standards and Technology (NIST)** 800-171 and 800-53 guidelines.

- **Regulatory standards** are very similar to government standards. These standards are specific to certain groups and regulated industries, such as healthcare, financial, pharmaceuticals, or factories. Within governments, these standards are generally assigned to a department to define and enforce the standards and laws for these industries. Examples are the **Financial Trade Commission (FTC)**, the **Occupational Safety and Health Association (OSHA)**, the **Environmental Protection Agency (EPA)**, and NIST. Within these regulatory groups, there are usually some industry standards to protect the privacy and the custodial handling of data. These can be through governmental or industry-based standards.

- **Industry standards** serve as the commonly accepted parameters within the various industry and technology sectors. They are established benchmarks that align with specific industry needs or government regulations, offering a framework for the utilization and safeguarding of technology and data. For instance, the **Payment Card Industry Data Security Standard (PCI-DSS)** is a critical standard for organizations handling credit card transactions, ensuring the security of payment data. Similarly, the **Health Insurance Portability and Accountability Act (HIPAA)** for the protection of **personal health information (PHI)**, among other requirements, is crucial for maintaining consistency, reliability, and security across various technological practices and data management protocols.

Understanding your company's business and geographic reach is necessary to determine the levels of compliance and the policies that need to be in place to monitor, manage, and maintain adherence to these standards.

Within Microsoft's cloud, there are tools and capabilities that allow you to govern your resources. For operational governance, **Azure Automation** can assist in patch management and scheduling updates. **Desired State Configuration (DSC)** provides standards for the baseline requirements for virtual machine deployment. **Azure Policy** is the foundation of enforcing compliance and auditing resources against compliance requirements. **Azure Blueprints** can be used to build a compliant deployment standard that can be used for the creation of new resources within Azure.

The **Microsoft Cloud Security Benchmark** is an initiative within Azure Policy that is enabled when you first create your Azure subscription. This allows you to begin to evaluate your company's resources against global best practices for security and compliance against industry-recognized frameworks. You can then begin to customize regulatory compliance based on specific requirements for government and industry compliance.

Once you have determined the requirements for compliance within your company, you can use **Microsoft Defender for Cloud** to evaluate the compliance of your Azure, on-premises, and multi-cloud infrastructure. The next section will discuss how you can accomplish these tasks.

Evaluating Infrastructure Compliance by Using Microsoft Defender for Cloud

Microsoft Defender for Cloud is Microsoft's **cloud security posture management (CSPM)** solution. Microsoft Defender for Cloud offers two options for CSPM within your Azure subscription:

- **Enhanced security off**, which is a free service
- **Enhanced security on**, which is a pay-as-you-go service based on the resources that you are monitoring

Though the free services within Microsoft Defender for Cloud provide a level of CSPM for your Azure infrastructure, you will need to change the settings for your subscription within Microsoft Defender for Cloud to enable Defender plans, such as Defender for Server, Defender for App Services, and Defender for Containers, for enhanced security features for advanced workloads and multi-cloud support.

A free subscription to Microsoft Defender for Cloud is turned on by default when you create an Azure tenant. The enhanced security features are turned on within the environment settings of Microsoft Defender for Cloud. Microsoft Defender for Cloud has many capabilities that a company can utilize for CSPM within Azure, hybrid, and multi-cloud infrastructures. *Figure 5.1* shows how to turn on and save the settings to add these Defender plans:

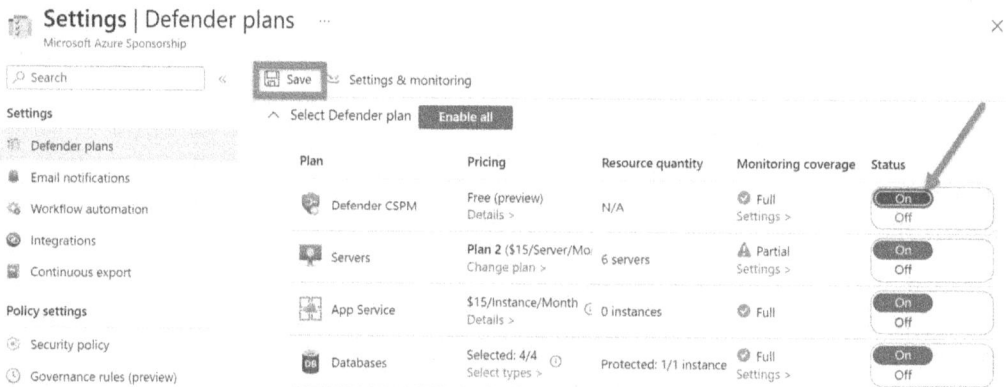

Figure 5.1: Turning on and saving Defender plans settings

As stated in the previous section, Microsoft Defender for Cloud policies are enabled by default with an Azure subscription.

Microsoft Defender for Cloud provides the following CSPM features to your Azure subscription:

- **Continuous assessment and security recommendations**: Azure Policy creates a policy named ASC Default in the subscription. This policy audits resources that are created within the Azure subscription on a continuous basis and provides recommendations for improvement within the Azure environment to be more secure based on best practices. The recommendations provided from these assessments are used as a guide for increasing your secure score and regulatory compliance. *Figure 5.2* shows these recommendations for controls that will improve the security posture and secure score:

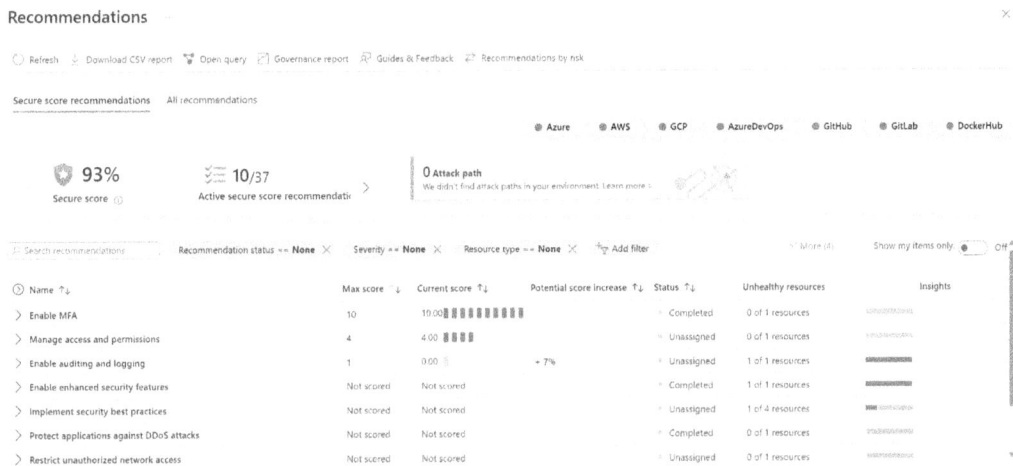

Figure 5.2: Security posture recommendations

Implementing these recommendations will improve your secure score and overall security posture.

- **Security posture**: The security posture provides the secure score for the subscription or cloud provider account. A secure score is a numerical value or percentage based on the assessment results and the level of best practices that are currently enabled. Implementing the controls that are recommended will have an increased effect on the secure score. The secure score is helpful for building a baseline for securing your resources and having a level of compliance. *Figure 5.3* shows the **Security posture** tile, which provides a quick overview of the Azure subscription and connected accounts from **Amazon Web Services** (**AWS**) and **Google Cloud Platform** (**GCP**):

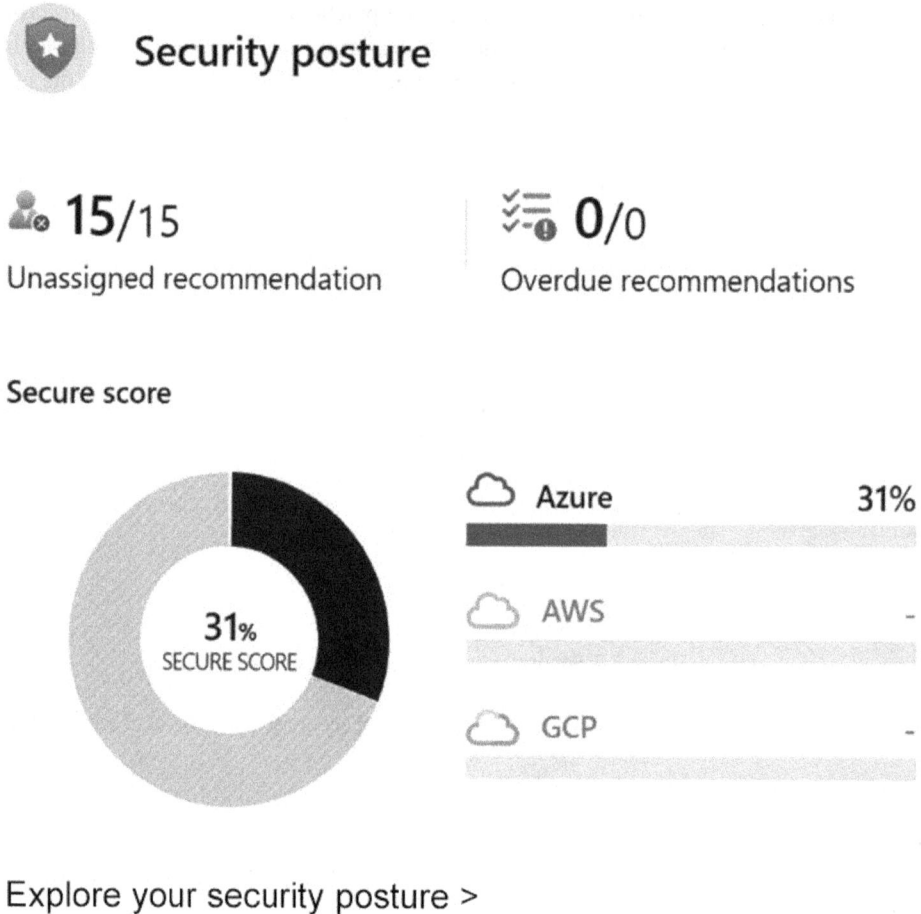

Figure 5.3: The Security posture overview tile

The **Security posture** dashboard is used for exploring and reviewing the recommendations for improving the security posture of your subscriptions. For a more detailed view against specific standards, you should use the **Regulatory compliance** dashboard.

- **Regulatory compliance**: This dashboard and its reports provide guidance based on regulatory standard initiatives. Defender plans that are enabled with enhanced security are required to use the **Regulatory compliance** dashboard. For companies that have regulatory compliance requirements, this dashboard will audit and assess your current controls and infrastructure against these requirements and help you prepare for a compliance audit. If a company does not require adherence to a particular standard, you still can measure their current controls against the Azure Security Benchmark. *Figure 5.4* shows the **Regulatory compliance** overview tile:

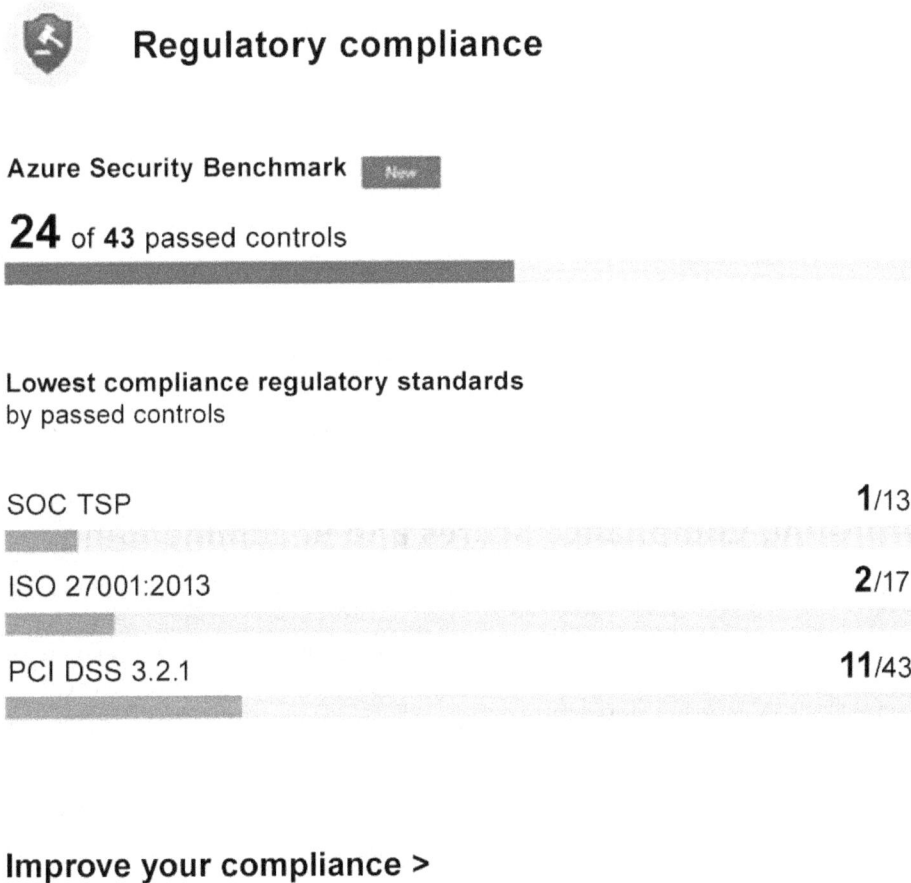

Regulatory compliance

Azure Security Benchmark   New

24 of **43** passed controls

Lowest compliance regulatory standards
by passed controls

SOC TSP   **1**/13

ISO 27001:2013   **2**/17

PCI DSS 3.2.1   **11**/43

Improve your compliance >

Figure 5.4: The Regulatory compliance overview tile

Selecting the **Improve your compliance** link will provide you with a breakdown of controls for that standard and the level of compliance for those resources. This is shown in *Figure 5.5*:

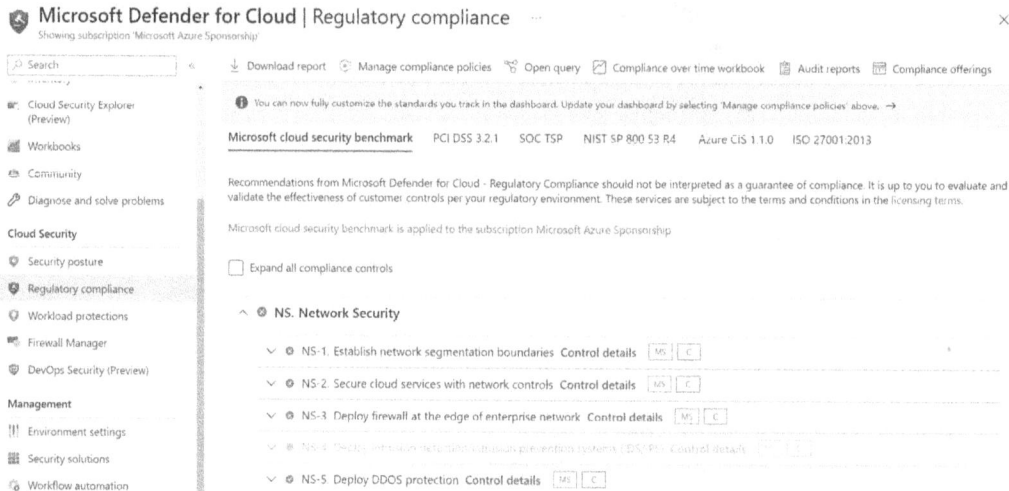

Figure 5.5: The Azure Security Benchmark compliance controls

The next section will discuss how to review these compliance scores and the actions that you can take within the recommendations to improve security and compliance.

Interpreting Compliance Scores and Recommending Actions to Resolve Issues or Improve Security

As stated in the previous section, Microsoft Defender for Cloud provides ways for you to review the compliance of Azure, AWS, and GCP resources. Using Azure Arc for non-Azure resources on-premises will also reveal information and recommendations on the security posture of those compute, network, and storage resources.

Reviewing the **Regulatory compliance** dashboard and the levels of compliance against resources will provide insights into company security. These insights and recommendations can be used to improve company security and prepare for compliance audits.

In Microsoft Defender for Cloud, you are provided with a view of the sections of the standard. Sections that are compliant are represented with a green checkmark, non-compliant sections are given a red "x," and areas that are not relevant to your current environment are grayed out. *Figure 5.6* provides a screenshot example of the Azure Security Benchmark sections of **Incident Response, Posture and Vulnerability Management, Endpoint Security, Backup and Recovery, DevOps Security**, and **Governance and Strategy**:

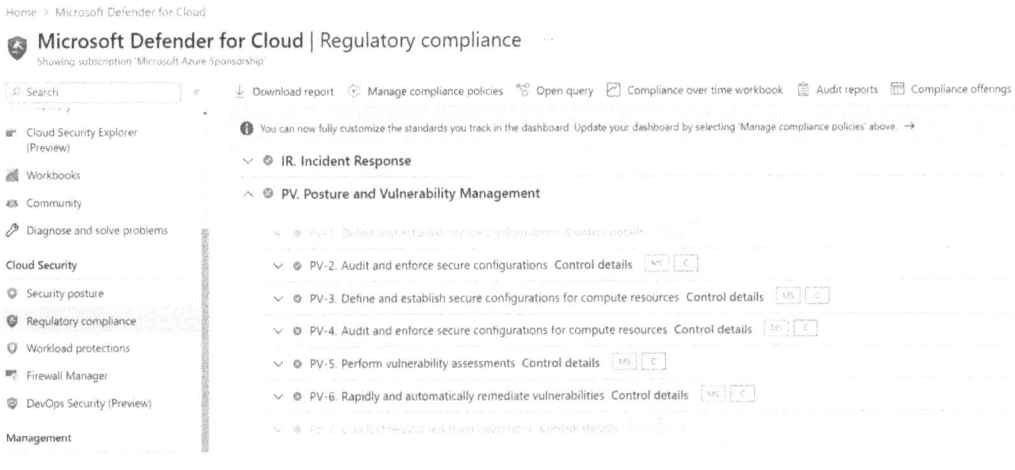

Figure 5.6: Regulatory compliance color representation

In the screenshot, you can see that **Incident Response** and **DevOps Security** are compliant with the Azure Security Benchmark based on the resources that are currently deployed within Azure. The **Governance and Strategy** section is not applicable to the current environment. **Endpoint Security**, **Backup and Recovery**, and **Posture and Vulnerability Management** are non-compliant and require additional action to be made compliant. Expanding the **Posture and Vulnerability Management** section shows the more detailed requirements, and you can review the additional details on the security controls required to ensure compliance. *Figure 5.7* shows an example of the more detailed recommendations:

Figure 5.7: Security control recommendations

These control recommendations can be implemented within your environment, or they can be reviewed and marked exempt or disabled, as shown in *Figure 5.8*:

SQL databases should have vulnerability findings resolved ...

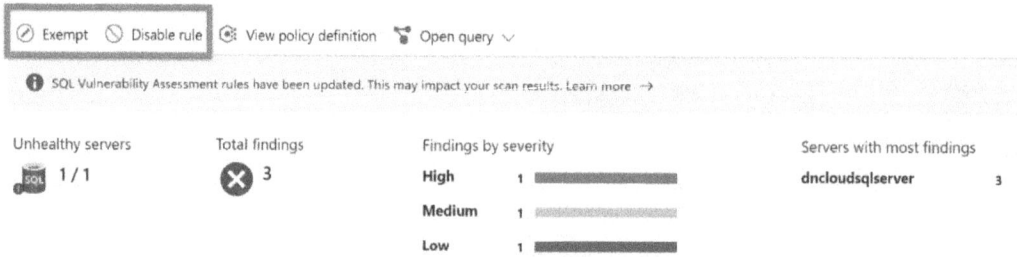

⊘ Exempt   ◯ Disable rule | ⚙ View policy definition   ⚓ Open query ∨

ⓘ SQL Vulnerability Assessment rules have been updated. This may impact your scan results. Learn more →

Unhealthy servers	Total findings	Findings by severity		Servers with most findings	
🗄 1 / 1	❌ 3	High	1 ▬▬▬▬▬▬	dncloudsqlserver	3
		Medium	1 ▬▬▬▬▬▬		
		Low	1 ▬▬▬▬▬▬		

Figure 5.8: Recommendation resolution

When resolving a recommendation by exempting or disabling the rule, you should have documentation and note the reason for the exemption. Microsoft Defender for Cloud will not recognize third-party controls that may provide company compliance without showing within the **Regulatory compliance** dashboard. Therefore, having documentation, justification, and notation within Microsoft Defender for Cloud is important.

In the next section, you will learn more about the integration of Azure Policy with Microsoft Defender for Cloud, and how to plan and design Azure Policy to govern implementation.

Designing an Implementation of Azure Policy

In the previous section, we discussed how to use Microsoft Defender for Cloud to determine the levels of compliance within the infrastructure. Addressing the recommendations takes you on a path to passing a regulatory audit. The ability to assess, monitor, and manage compliance with these standards and other standards within Azure is done using Azure Policy. *Figure 5.9* shows the workflow of a policy within Azure:

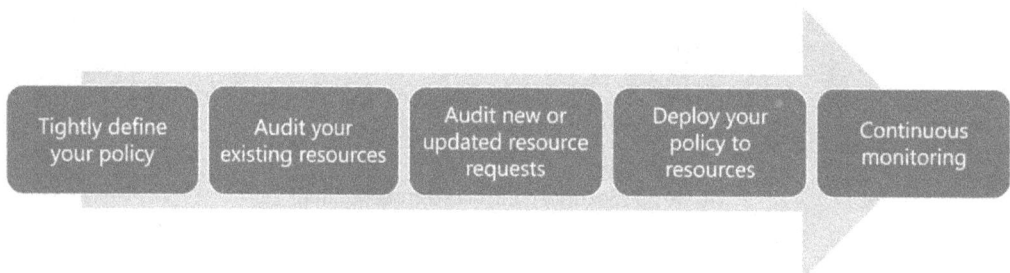

Tightly define your policy → Audit your existing resources → Audit new or updated resource requests → Deploy your policy to resources → Continuous monitoring

Figure 5.9: Azure Policy workflow

Azure Policy is used to create definitions of governance parameters that meet a company's standards. These policies are then continuously evaluated for compliance and any changes within the environment. Azure Policy definitions are not only used for regulatory compliance to standards; they can also be used to create cost and size parameters on resources and to ensure that logging and monitoring tools are included upon deployment.

Azure Policy is a service that allows you to create and manage policies for your Azure resources. These policies enforce rules, ensuring compliance with corporate regulations and standards, security, and cost parameters. You define policy rules (JSON format), assign them to specific scopes (e.g., subscriptions and resource groups), and evaluate resources against these rules. If a resource violates a policy, it fails compliance checks, prompting remediation. Azure Policy extends governance across cloud providers through Azure Arc. Azure Policy is proactive to new resources being deployed and can also audit existing resources. The continuous assessment of resources maintains levels of compliance, as shown in *Figure 5.10*:

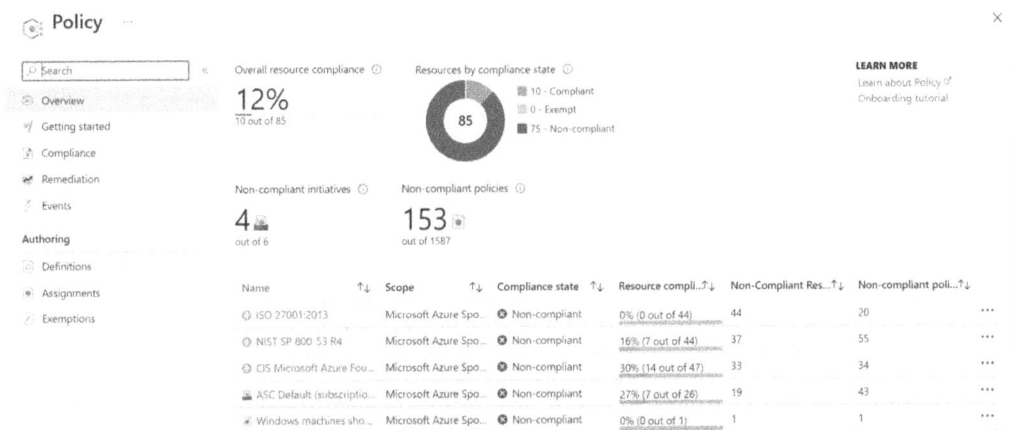

Figure 5.10: Policy compliance overview

Figure 5.10 shows some of the standards and regulations that are being governed by Azure Policy. If you look at the **Non-compliant policies** column, you will see that there is more than one policy that creates the level of regulatory and standards compliance. This group of policies is called an initiative within Azure. These initiatives are made up of two or more policy definitions.

Policies are nothing more than a definition of parameters that a resource must adhere to when deployed and used within Azure. Azure Policy has a pre-built policy and initiative definitions for different resource categories. You can also create your own policy or initiative definition to meet your company standards and guidelines. *Figure 5.11* shows some of the built-in policies and how you can select + **Policy definition** or + **Initiative definition**:

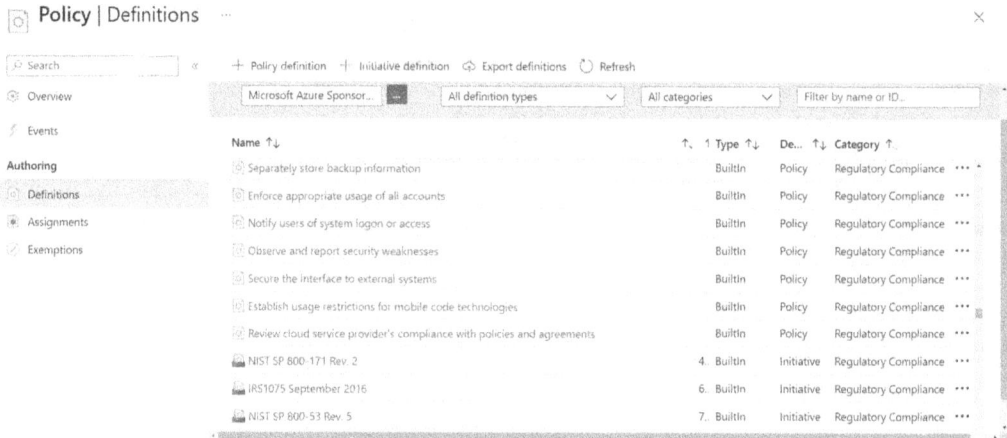

Figure 5.11: Built-in policies and + Policy definition and + Initiative definition

A good practice when using policies is to have a plan and review the governance standards with the company stakeholders. This should include the security, finance, and development teams to understand the various needs of each group to protect, control, and govern your Azure resources. To avoid frustration, these policies should be documented and published as deployment guidelines within Azure. If you have tight policies that are communicated, then the contributors creating resources will not have failed deployments if these published policies are followed. Suppose the development team wants to deploy resources in the US East region, but the policy is to only allow deployments to the US West region. So, when the US East region is selected, because of the policy, it will show in the portal that the wrong region has been selected.

The next section will discuss how to address data residency within your cloud resources.

Designing for Data Residency Requirements

The cloud is a global resource that expands your company footprint well beyond the native country of origin. However, with these expanded capabilities to create resources to get closer to your users, there is a challenge in regulations as it pertains to data sovereignty and data residency.

In light of stringent government and privacy mandates, certain nations enforce rigorous protocols for data management. Data residency relates to the specific geographical location where data is stored and processed. Conversely, data sovereignty encompasses the legal framework imposed by national or local authorities that governs the treatment of data. Compliance with these regulations ensures that data handling aligns with the prescribed standards of the jurisdiction in question. When designing resources within the cloud, you need to understand the local jurisdiction for the regions where these resources are deployed. Not adhering to local standards could create a level of legal liability. In addition, if you have data resident within a region that is connected openly to another region's data, the local government could be able to seize the data within the other region. Think of a scenario where your company has recently expanded its operations to handle credit card transactions and has also started doing business in Germany. The company is concerned about adhering to regulatory compliance standards, data residency requirements, and privacy regulations. The solution could be using Microsoft Defender for Cloud and implementing Azure Policy to address the concern.

Proper planning and understanding of local and country laws and regulations can help prevent future legal issues when it comes to a legal hold or eDiscovery case. Azure Policy can be used to assist in governing data sovereignty. If you are creating resources within specific regions, you should have policies in place that have location parameters to only deploy within those regions. It is a good practice to design your resource groups by region and then create a policy for each of those resource groups to use only that country's regions. *Figure 5.12* shows some of the built-in policies for allowed locations that can be used to design for data residency:

Name ↑↓	Latest version (preview) ↑↓	Definition location ↑↓	Policies ↑↓	Type ↑↓	Definition type ↑↓	Category ↑↓
Allowed locations for resource groups	1.0.0			Builtin	Policy	General
Configure subscriptions to set up preview features	1.0.1			Builtin	Policy	General
Allowed locations	1.0.0			Builtin	Policy	General
Audit usage of custom RBAC roles	1.0.1			Builtin	Policy	General
Allowed resource types	1.0.0			Builtin	Policy	General
Do not allow deletion of resource types	1.0.1			Builtin	Policy	General
Not allowed resource types	2.0.0			Builtin	Policy	General
Do Not Allow MCPP resources	1.0.0			Builtin	Policy	General
Do Not Allow M365 resources	1.0.0			Builtin	Policy	General

Figure 5.12: Allowed location policies

Data residency and sovereignty are important for business as well as personal data usage privacy. In the next section, we will further discuss how to design security and governance around privacy requirements.

To assess the data residency, data classification, information protection, privacy, insider risk management, information barriers, and data loss prevention for the whole tenant, it is advisable to use Microsoft Purview from this link: https://purview.microsoft.com.

Translating Privacy Requirements into Requirements for Security Solutions

As stated in the previous section, data residency is an important aspect of protecting the privacy of data, both business and personal. In this section, we will look more closely at some of the ways that you can protect the privacy of data, identities, and resources within Microsoft 365 and Azure.

Various privacy standards exist to safeguard sensitive data, including PII, PHI, and PCI-DSS. These standards are established by global (ISO/IEC 27018), regional (GDPR), and local (HIPAA) authorities. Their common objective is to prevent sensitive information from falling into unauthorized personnel's hands.

These attackers, or bad actors, attempt to gain access to data that is easily accessible and in plain text. It is the responsibility of the cybersecurity architect to design a solution that maintains levels of encryption throughout the storage, transportation, and use of the data. Most cloud providers have at-rest data protection with encryption at rest. If this data is encrypted at rest and the location becomes exposed to an attacker, they will not be able to decrypt the data to read it, since they do not have the proper credentials. Protecting user identities behind multi-factor authentication provides that layer of protection to not allow the attacker access to the encrypted data.

Microsoft Azure has multiple ways to encrypt data, depending on the service that is being used:

- Azure **Storage Service Encryption** (**SSE**) is used with Azure storage accounts and is enabled by default to encrypt data at rest. Settings within storage accounts can also be set to only use HTTPS to access a storage account, which provides encryption in transit.

- **Azure Disk Encryption** (**ADE**) is used to encrypt virtual machines and their storage at the OS and disk levels. Similar to encryption on a Windows device, ADE uses BitLocker for Windows devices. For Linux devices, `dm-crypt` is used for encryption.

- Database encryption for SQL databases uses **transparent data encryption** (**TDE**) to encrypt the databases at rest. Dynamic data masking can also be used to prevent data exposure to unauthorized users for data such as credit card numbers and social security numbers.

- Azure Key Vault is the service within Azure to manage encryption keys outside of a particular service. Some regulatory standards require a separation of duties between the keys and the services. Azure Key Vault provides this separation.

In the preceding list, it was mentioned that with Azure Storage, you can set data in transit to only be transferred using HTTPS. This provides the encryption of data in transit utilizing SSL/TLS protocols; TLS 1.2 or later is the current standard. This is the minimum requirement to protect the privacy of data in transit. When possible, a more secure communication channel such as a **virtual private network** (**VPN**) from the on-premises network to the cloud services should be used. Azure can also protect the transmission of data with Azure Application Gateway and Azure Front Door, which use a **web application firewall** (**WAF**) to protect against common security threats.

When discussing data privacy within Microsoft, this all falls under governance and compliance. Microsoft Purview provides complete insight into data from Azure and Microsoft 365 resources as well as other cloud, SaaS, and on-premises resources. Utilizing the solutions in Microsoft Purview provides the full mapping, monitoring, protection, and governance of data across the infrastructure. *Figure 5.13* provides a diagram that shows the services within and outside of Microsoft that can be governed and protected:

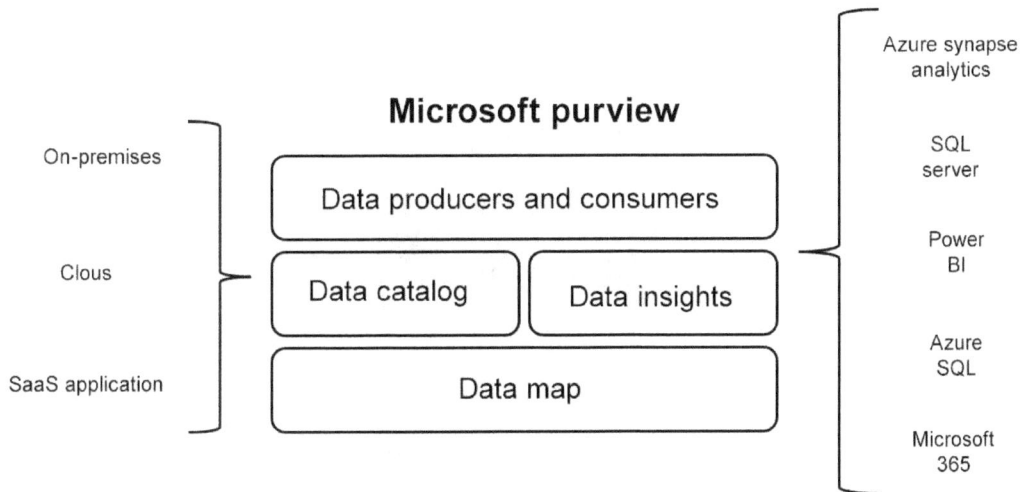

Figure 5.13: Microsoft Purview for data privacy governance

Microsoft Purview is a unified governance solution that automates data discovery and management across your data estate. At its core, the Purview Data Map acts as a **platform-as-a-service (PaaS)** component, maintaining an up-to-date map of assets and their metadata. To build this map, you register and scan your data sources. Collections within Purview allow you to organize and manage data sources, scans, and assets hierarchically. By defining collections, you can tailor a custom model of your data landscape based on your organization's governance needs. Collections also serve as security boundaries, ensuring least-privilege access to metadata. With Purview, you can delegate ownership, search assets efficiently, and enforce compliance across hybrid, on-premises, and multi-cloud environments.

> **Note**
>
> More information on Microsoft Purview can be found at this link: `https://learn.microsoft.com/en-us/purview/purview`.

The next section will provide a case study that summarizes the process of evaluating a strategy for regulatory compliance, as discussed in this chapter.

Case Study – Designing for Regulatory Compliance

Apply what you learned in this chapter by completing the case study on the accompanying online platform. In this case study, you will be given a company scenario and asked to complete several tasks to ensure the company meets regulatory compliance standards.

To access the case study, visit the following link or scan the QR code.

Link to the case study: `https://packt.link/SC100-E2-CaseStudy_Chapter5`

QR code:

Figure 5.14: QR code to access case study for Chapter 5

Summary

In this chapter, we discussed how to design and evaluate a strategy for regulatory compliance, including the continuous assessment of resources within Microsoft Defender for Cloud for multi-cloud and hybrid infrastructures, governing our Azure environment with Azure Policy, and maintaining and adhering to privacy and data residency requirements. You will have gained a comprehensive understanding of how to design and implement regulatory compliance strategies, evaluate and manage compliance using Microsoft Defender for Cloud and Azure Policy, and address data residency and privacy requirements. This knowledge is crucial for both practical application and exam preparation.

In the next chapter, you will learn how to evaluate security posture and recommend technical strategies to manage risk. This involves interpreting compliance scores and recommending actions to resolve issues or improve security. Reviewing the **Regulatory compliance** dashboard and the levels of compliance against resources will provide insights into company security. These insights and recommendations can be used to improve company security and prepare for compliance audits.

Exam Readiness Drill – Chapter Review Section

Apart from mastering key concepts, strong test-taking skills under time pressure are essential for acing your certification exam. That's why developing these abilities early in your learning journey is critical.

Exam readiness drills, using the free online practice resources provided with this book, help you progressively improve your time management and test-taking skills while reinforcing the key concepts you've learned.

How to Get Started

1. Open the link or scan the QR code at the bottom of this page.

2. If you have unlocked the practice resources already, log in to your registered account. If you haven't, follow the instructions in *Chapter 11* and come back to this page.

3. Once you have logged in, click the **START** button to start a quiz.

We recommend attempting a quiz multiple times till you're able to answer most of the questions correctly and well within the time limit.

You can use the following practice template to help you plan your attempts:

Working On Accuracy		
Attempt	Target	Time Limit
Attempt 1	40% or more	Till the timer runs out
Attempt 2	60% or more	Till the timer runs out
Attempt 3	75% or more	Till the timer runs out
Working On Timing		
Attempt 4	75% or more	1 minute before time limit
Attempt 5	75% or more	2 minutes before time limit
Attempt 6	75% or more	3 minutes before time limit

The above drill is just an example. Design your drills based on your own goals and make the most of the online quizzes accompanying this book.

First time accessing the online resources? 🔒

You'll need to unlock them through a one-time process. **Head to** *Chapter 11* **for instructions.**

Open Quiz	
https://packt.link/SC100_CH05	
Or scan this QR code →	

Evaluate Security Posture and Recommend Technical Strategies to Manage Risk

The previous chapter discussed how to design security and governance strategies based on regulatory compliance requirements within your company. This included how to utilize Microsoft Defender for Cloud and Azure Policy to evaluate and govern your company resources.

In this chapter, you will explore various aspects of evaluating and managing the security posture of your cloud infrastructure. This chapter will cover the domain **Design security solutions for infrastructure** in the SC-100 exam guide. You will begin by understanding the importance of using benchmarks such as the Azure security benchmarks and ISO 27001 to assess your security posture. Next, you will explore the capabilities of Microsoft Defender for Cloud, which provides tools for continuous assessment and security recommendations. You will also cover the concept of secure scores and how they can help you identify areas for improvement in your security posture. Additionally, you will examine the security of cloud workloads and the role of **extended detection and response (XDR)** capabilities in protecting your infrastructure.

In this chapter, you are going to cover the following main topics:

- Evaluating the security posture using benchmarks, including Azure security benchmarks, ISO 27001, and more
- Evaluating the security posture using Microsoft Defender for Cloud
- Evaluating the security posture using Secure Score
- Evaluating the security posture of cloud workloads
- Designing security for an Azure landing zone
- Interpreting technical threat intelligence and recommending risk mitigations

- Recommending security capabilities or controls to mitigate identified risks

- Designing a solution for evaluating the security posture of internet-facing assets using Microsoft **Defender External Attack Surface Management (Defender EASM)**

By the end of this chapter, you will have a comprehensive understanding of how to evaluate and enhance the security posture of your cloud environment.

Evaluating the Security Posture Using Benchmarks

This section will focus on the various benchmarks for evaluating the security posture within Microsoft Defender for Cloud and Azure Policy initiatives. Before evaluating these benchmarks further, however, you should understand what is meant by **cloud security posture management (CSPM)**.

CSPM involves monitoring and managing security defenses to audit, assess, and identify potential vulnerabilities and threats within our infrastructure. By maintaining this continuous process, you can proactively address the possibility of attacks before they occur, ensuring diligence and adaptability in the face of evolving threats and cloud environments. *Figure 6.1* shows the continuous process of CSPM:

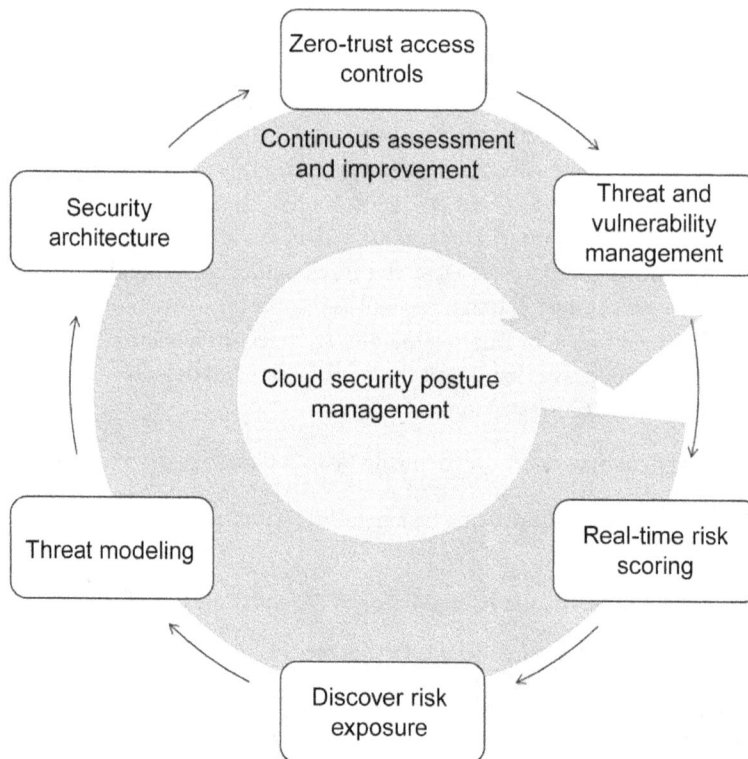

Figure 6.1: CSPM: continuous assessment and improvement

Each of the areas shown in *Figure 6.1* assists in adding new security controls and improving the cloud security posture. A strong CSPM solution evaluates and provides the following features and characteristics:

- **Zero-trust access control**: Zero-trust access control involves considering the active threat level when making access control decisions

- **Threat and vulnerability management**: Threat and vulnerability management involves providing a holistic view of the organization's attack surface and risk, and integrating this information into decision-making

- **Technical policy**: To audit and enforce the organization's standards and policies on technical systems, guardrails are used

- **Real-time risk scoring**: Real-time risk scoring provides visibility into the top risks

- **Discover sharing risks**: To understand the data exposure of enterprise intellectual property on sanctioned and unsanctioned cloud services

- **Advanced threat protection features**: By using the Defender plans for compute, data, and service layers, workloads are protected with advanced alerts to respond to potential threats

Within Microsoft Azure, Microsoft Defender for Cloud (formerly Azure Security Center) provides CSPM for hybrid and multi-cloud infrastructure resources.

Before continuing, let's clarify what is meant by hybrid and multi-cloud infrastructures:

- Hybrid infrastructures are a combination of cloud infrastructures, such as Microsoft Azure, and an on-premises data center. *Figure 6.2* shows a hybrid infrastructure environment – on-premises and Microsoft. This on-premises infrastructure may be a customer's data center or a co-located data center that they are sharing with other customers:

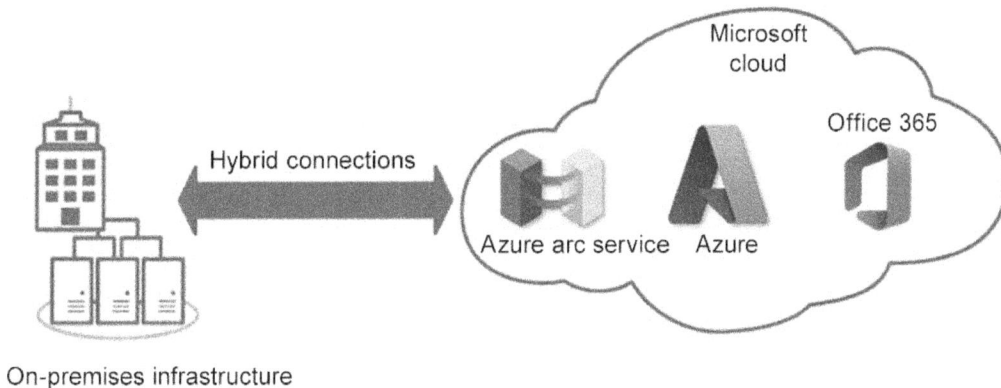

Figure 6.2: Hybrid infrastructure diagram

Some companies may utilize more than one cloud provider with Azure. This use of additional cloud providers is known as a multi-cloud infrastructure:

- Multi-cloud infrastructures are more complex and, at times, misinterpreted. Multi-cloud is an infrastructure that combines more than one cloud provider. Typically used cloud providers include Azure, AWS, and GCP, though there are others, including Alibaba, Oracle, and IBM. A multi-cloud infrastructure could also include the hybrid component of on-premises infrastructure – for example, Azure plus AWS or GCP is a multi-cloud infrastructure, as is Azure combined with AWS and on-premises.

AWS and GCP are the current cloud providers whose accounts and projects can be directly integrated into Microsoft Defender for Cloud. If you are utilizing any other cloud providers for virtual machine instances, they can be connected by utilizing the hybrid approach with Azure Arc. *Figure 6.3* shows AWS and GCP added as a multi-cloud infrastructure:

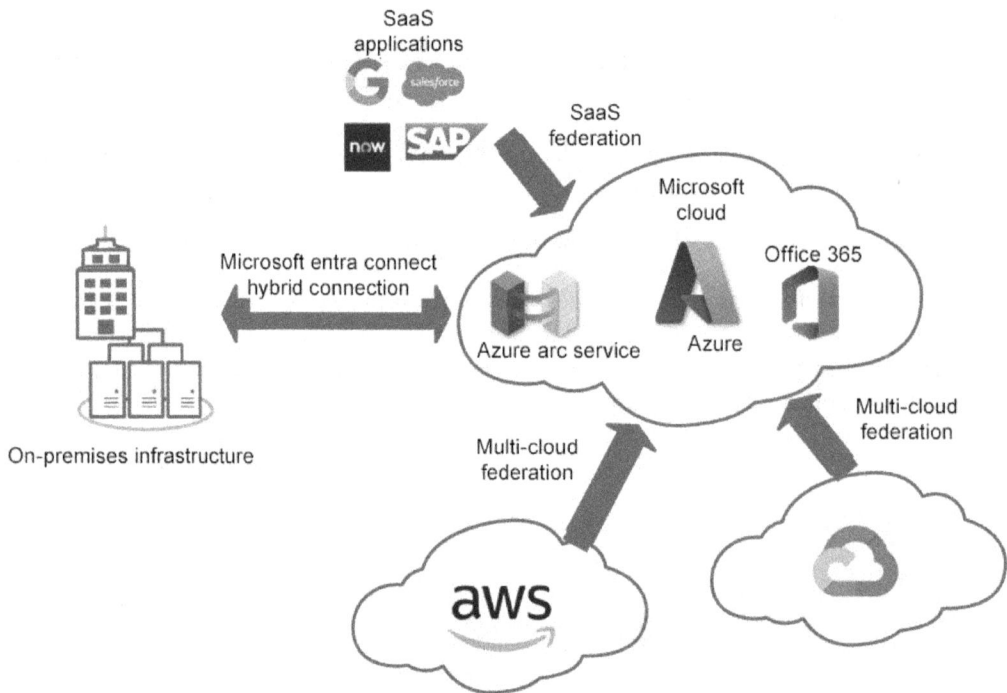

Figure 6.3: Multi-cloud infrastructure diagram

Microsoft Defender for Cloud provides CSPM and threat and vulnerability protection across these infrastructures. Providing this level of CSPM can be accomplished by utilizing benchmark standards provided within Microsoft Defender for Cloud. These benchmarks allow you to evaluate the current level of compliance for your infrastructure and make remediation decisions based on targeted recommendations.

The Microsoft cloud security benchmark was discussed briefly in *Chapter 5, Design a Regulatory Compliance Strategy*. This is an initiative within Azure Policy that is enabled when you first create your Azure subscription. The Azure Security Benchmark is applied automatically and reviews your infrastructure. This allows you to begin to evaluate your company's resources against global best practices for security and compliance against industry-recognized frameworks. You can then begin to customize it based on specific requirements based on government and industry compliance.

When utilizing the enhanced security features included with Microsoft Defender plans, additional compliance initiatives become available. These include PCI-DSS 3.2.1, ISO 27001:2013, NIST SP 800-53, and others. However, it's important to note that not all compliance initiatives review security posture across multi-cloud infrastructures. For instance, AWS resources adhere to AWS CIS 1.2.0 and AWS PCI DSS 3.2.1 regulatory standards, while Google Cloud Platform assigns the GCP default security baseline and offers compliance reviews for GCP CIS 1.1.0, GCP CIS 1.2.0, GCP ISO 27001, GCP NIST 800-53, and PCI DSS 3.2.1. Understanding these nuances is crucial for maintaining a robust security posture across multi-cloud environments.

> **Note**
>
> For the full list of regulatory and benchmark standards available, it is recommended to go through this link as it is important for the exam: `https://learn.microsoft.com/en-us/azure/defender-for-cloud/update-regulatory-compliance-packages`.

Let's look at how you can evaluate and make adjustments to your CSPM by utilizing the Azure Security Benchmark by walking through the **Regulatory compliance** section of Microsoft Defender for Cloud.

Think of a scenario – for example, Contoso, a financial services company, faced a significant challenge when confidential client information was exposed due to a misconfigured network security group. This breach led to financial penalties and damaged their reputation. To address this, Contoso leveraged Microsoft Defender for Cloud to enhance its security posture. By implementing continuous assessment and security recommendations, they identified and remediated misconfigurations in real time. The use of Secure Score helped prioritize critical security controls, such as enabling **multi-factor authentication (MFA)** and network hardening. Additionally, Contoso utilized threat and vulnerability management to gain a holistic view of their attack surface and integrate this information into decision-making. This proactive approach allowed them to address potential vulnerabilities before they could be exploited. As a result, Contoso resolved its immediate security issues, established a proactive defense mechanism, and restored its clients' trust, positioning itself as a leader in data security within the financial industry.

To do this in the Azure portal, the following steps are taken:

1. In the Azure portal (`https://portal.azure.com/`), search for and navigate to **Microsoft Defender for Cloud**.

2. Navigate to **Regulatory compliance** under **Cloud Security** on the menu, as shown in *Figure 6.4*:

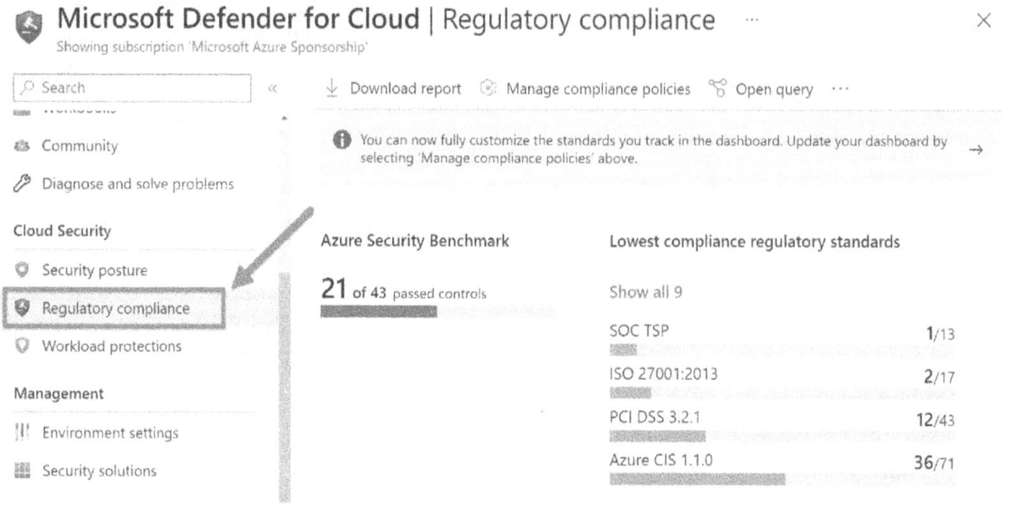

Figure 6.4: Regulatory compliance in the Cloud Security menu

3. If this section is not available to you, you will need to go to your subscription in **Environment settings** to turn on the Microsoft Defender plans. These steps will be reviewed in the *Evaluating the Security Posture of Cloud Workloads* section.

4. Next, scroll down within the **Regulatory compliance** overview dashboard and select **Azure Security Benchmark**. This is usually the default selection as the first tab. Note the tabs for the additional **Regulatory compliance** standards that are active within your subscription, as shown in *Figure 6.5*:

Figure 6.5: List of regulatory compliance options

5. Scroll down to view the compliance controls for the Azure Security Benchmark. These controls match the documentation for the compliance benchmark or standard that you are viewing. Documentation for these controls for the Azure Security Benchmark can be found here: `https://learn.microsoft.com/en-us/security/benchmark/azure/ overview`. *Figure 6.6* shows a list of these controls in Microsoft Defender for Cloud:

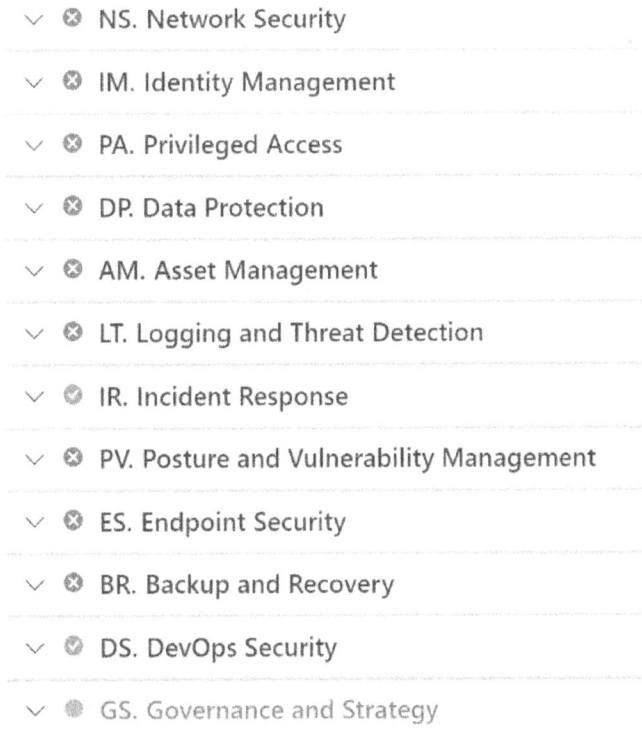

> ∨ ⊗ NS. Network Security
>
> ∨ ⊗ IM. Identity Management
>
> ∨ ⊗ PA. Privileged Access
>
> ∨ ⊗ DP. Data Protection
>
> ∨ ⊗ AM. Asset Management
>
> ∨ ⊗ LT. Logging and Threat Detection
>
> ∨ ⊘ IR. Incident Response
>
> ∨ ⊗ PV. Posture and Vulnerability Management
>
> ∨ ⊗ ES. Endpoint Security
>
> ∨ ⊗ BR. Backup and Recovery
>
> ∨ ⊘ DS. DevOps Security
>
> ∨ ⊛ GS. Governance and Strategy

Figure 6.6: List of Azure Security Benchmark controls

6. These control sections provide a quick review of your level of compliance and security posture within the current infrastructure:

- Green checkmarks show that you are fully compliant with those controls
- Red checkmarks show that you are missing controls
- If the control is gray, you currently do not have any applicable areas of your environment that apply to this control

7. Select one of the controls with a red checkmark – for example, **NS. Network Security** – as shown in *Figure 6.7*:

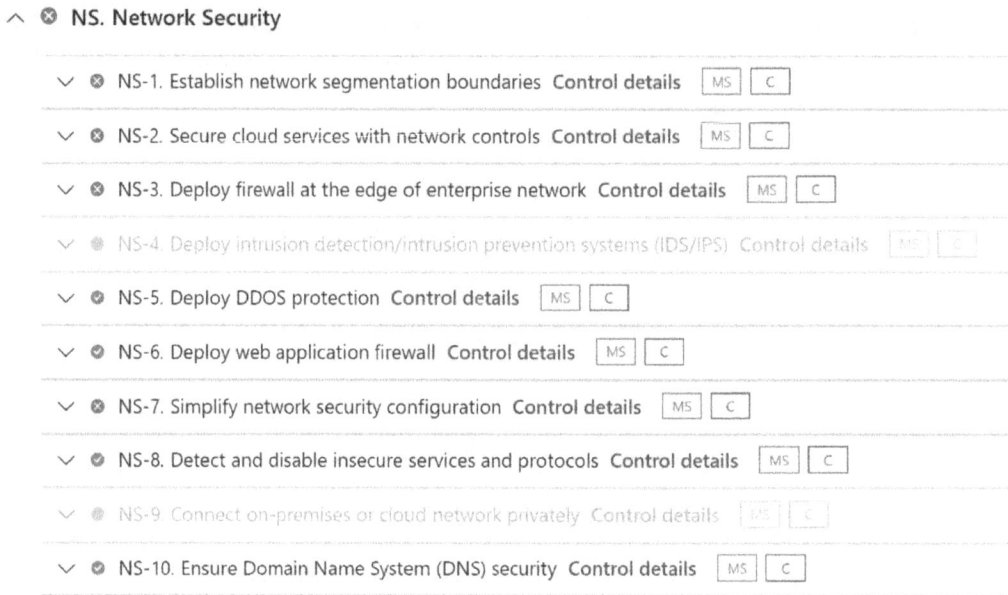

∧ ⊗ **NS. Network Security**

    ∨ ⊗ NS-1. Establish network segmentation boundaries **Control details** `MS` `C`

    ∨ ⊗ NS-2. Secure cloud services with network controls **Control details** `MS` `C`

    ∨ ⊗ NS-3. Deploy firewall at the edge of enterprise network **Control details** `MS` `C`

    ∨ ● NS-4. Deploy intrusion detection/intrusion prevention systems (IDS/IPS) Control details `MS` `C`

    ∨ ⊘ NS-5. Deploy DDOS protection **Control details** `MS` `C`

    ∨ ⊘ NS-6. Deploy web application firewall **Control details** `MS` `C`

    ∨ ⊗ NS-7. Simplify network security configuration **Control details** `MS` `C`

    ∨ ⊘ NS-8. Detect and disable insecure services and protocols **Control details** `MS` `C`

    ∨ ● NS-9. Connect on-premises or cloud network privately Control details `MS` `C`

    ∨ ⊘ NS-10. Ensure Domain Name System (DNS) security **Control details** `MS` `C`

Figure 6.7: Network security controls

8. The **MS** and **C** boxes denote the responsibility of Microsoft and the customer, respectively. Selecting the dropdown of the control will provide the resources and recommendations to further evaluate what needs to be done from a customer responsibility perspective to improve the security posture, as shown in *Figure 6.8*:

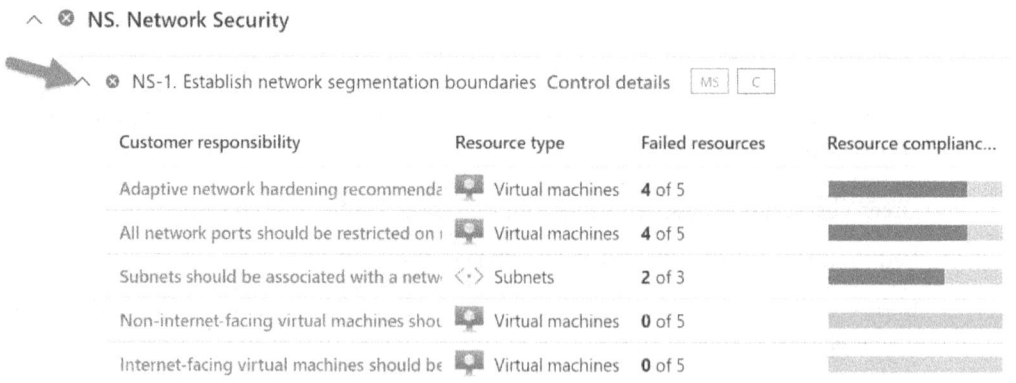

∧ ⊗ **NS. Network Security**

    ⊗ NS-1. Establish network segmentation boundaries Control details `MS` `C`

Customer responsibility	Resource type	Failed resources	Resource complianc...
Adaptive network hardening recommenda	Virtual machines	4 of 5	████████████░░░
All network ports should be restricted on	Virtual machines	4 of 5	████████████░░░
Subnets should be associated with a netw	Subnets	2 of 3	████████░░░░░░░
Non-internet-facing virtual machines shou	Virtual machines	0 of 5	░░░░░░░░░░░░░░░
Internet-facing virtual machines should be	Virtual machines	0 of 5	░░░░░░░░░░░░░░░

Figure 6.8: Recommended security controls to improve security posture

As a cybersecurity architect who is evaluating a customer's environment to provide recommendations for increasing security posture, you should review these recommendations and provide guidance on how to implement these to the various administration and architecture teams. Selecting these recommendations shows the non-compliant and compliant resources, as well as the steps to remediate these resources for security control compliance. These steps can be repeated for other compliance standards with the same control headings that match the documentation for the standards. *Figure 6.9* shows the list for PCI-DSS 3.2.1:

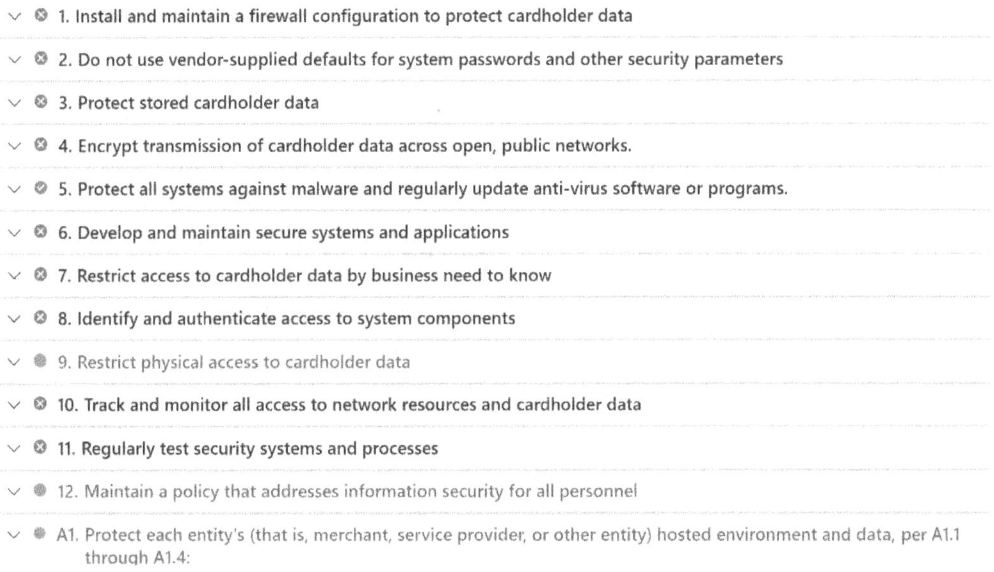

∨ ⊗ 1. Install and maintain a firewall configuration to protect cardholder data

∨ ⊗ 2. Do not use vendor-supplied defaults for system passwords and other security parameters

∨ ⊗ 3. Protect stored cardholder data

∨ ⊗ 4. Encrypt transmission of cardholder data across open, public networks.

∨ ⊘ 5. Protect all systems against malware and regularly update anti-virus software or programs.

∨ ⊗ 6. Develop and maintain secure systems and applications

∨ ⊗ 7. Restrict access to cardholder data by business need to know

∨ ⊗ 8. Identify and authenticate access to system components

∨ ◉ 9. Restrict physical access to cardholder data

∨ ⊗ 10. Track and monitor all access to network resources and cardholder data

∨ ⊗ 11. Regularly test security systems and processes

∨ ◉ 12. Maintain a policy that addresses information security for all personnel

∨ ◉ A1. Protect each entity's (that is, merchant, service provider, or other entity) hosted environment and data, per A1.1 through A1.4:

Figure 6.9: PCI-DSS 3.2.1 list of security control requirements

Evaluating the security posture by evaluating regulatory compliance within Microsoft Defender for Cloud is helpful to prepare your company for an upcoming audit, as well as to verify and report to executives that you are doing your due diligence and due care to protect the environment against identity, device, and data exposure. However, this is not the only area where you can evaluate your cloud security posture.

In the next section, you will learn how to use the **Security posture** area for Azure, hybrid, and multi-CSPM.

Evaluating the Security Posture Using Microsoft Defender for Cloud

Microsoft Defender for Cloud provides features and capabilities for understanding your security based on best practices and protecting workloads from vulnerabilities and threats. Microsoft Defender for Cloud also has integrations with tools for automation, support, remediation, and security operations. *Figure 6.10* shows how Microsoft Defender for Cloud integrates with other Microsoft and third-party tools for a fully integrated solution for CSPM, **Cloud Workload Protection Platform** (**CWPP**), investigation, remediation, and security operations:

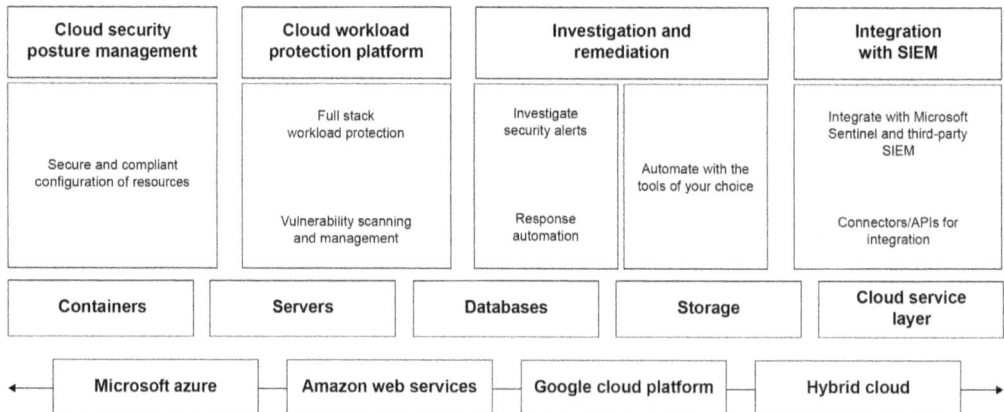

Figure 6.10: Microsoft Defender for Cloud capabilities

Microsoft Defender for Cloud has many capabilities that a company can utilize for CSPM within Azure, hybrid, and multi-cloud infrastructures.

Defender for Cloud provides the following CSPM features for your Azure subscription:

- **Continuous assessment and security recommendations**: Azure Policy creates a policy named ASC Default on the subscription. This policy is based on the Azure Security Benchmark (v3 at the time of writing). The Azure Security Benchmark initiative audits resources that are created within the Azure subscription continuously and provides recommendations for improvement within the Azure environment to be more secure based on best practices.

- **Secure Score**: The secure score is a numerical value or percentage based on the assessment results and the level of best practices that are currently enabled. Implementing the controls that are recommended will have an increased effect on the secure score.

The continuous assessment results of Azure resources are used to provide the secure score. The secure score consists of the following components:

- **Current score**: The score based on the controls that currently contribute to the total score.

- **Max score**: The points you can gain by implementing all recommendations for a control.

- **Potential increase**: The score increases if you remediate all recommendations.

- **Improvement actions**: Recommendations for controls that can increase your secure score. Quick Fix applies these immediately.

The components for continuous assessment help with reviewing the Azure infrastructure and determining actions that can be applied to better protect your subscription against vulnerabilities and threats. This continuous assessment can be reviewed and evaluated within the **Security posture** area under the **Cloud Security** menu, as shown in *Figure 6.11*:

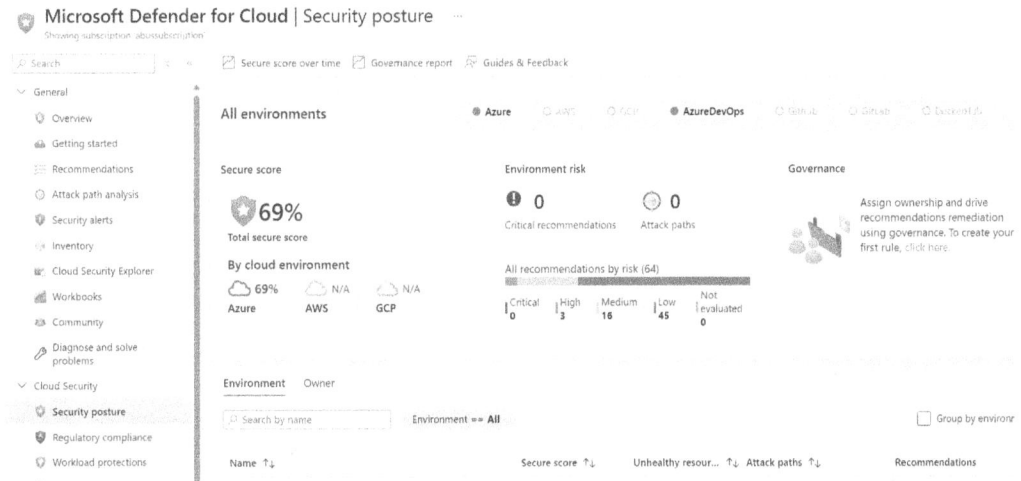

Figure 6.11: Security posture in the Cloud Security menu

The review of the **Security posture** area within Microsoft Defender for Cloud is based on the Azure Security Benchmark (currently v3). Unlike the review of the Azure Security Benchmark with regulatory compliance, recommendations within the **Security posture** area are not organized in a way that is aligned with the documentation of the standard. The recommendations are aligned more toward increasing your secure score through enabling recommended control because it provides a quantifiable measure of your security posture. The secure score helps you prioritize the most critical security controls that need to be implemented to protect your cloud environment.

To view these recommendations, scroll down within the **Security posture** area and select **View recommendations** within the subscription, as shown in *Figure 6.12*:

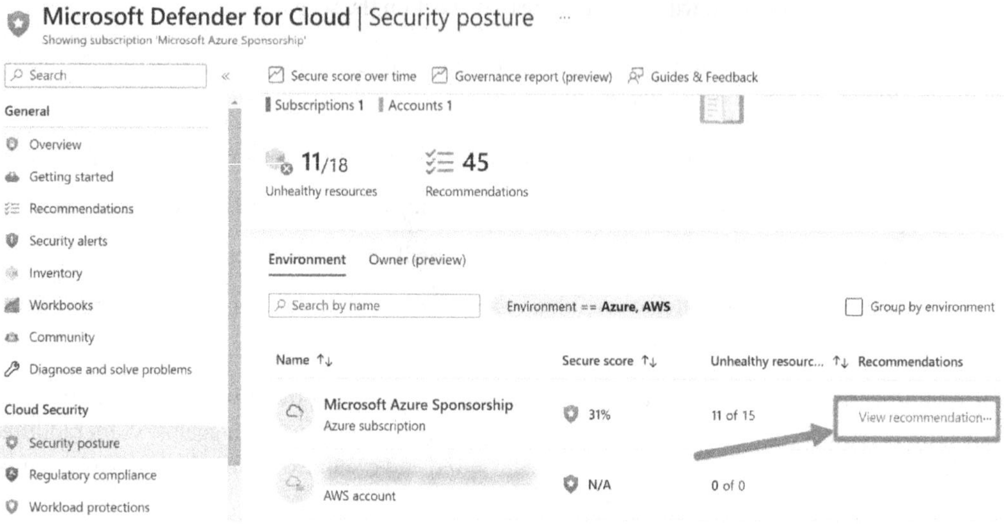

Figure 6.12: Subscription security posture recommendations

The next section will discuss how to evaluate these recommendations to increase the secure score and strengthen your security posture.

Evaluating the Security Posture by Using Secure Score

As stated in the previous section, Microsoft Defender for Cloud provides ways for you to evaluate current security controls and recommendations for additional controls to improve your environment's security posture. The overall security posture of the environment is given a percentage or number value known as the secure score.

Reviewing and evaluating the security posture recommendations allows you to increase your secure score. When selecting the **View recommendations** option, as shown in *Figure 6.12*, you will be taken to the list of recommended security controls. By default, these are listed by the highest impact on the secure score, as shown in *Figure 6.13*:

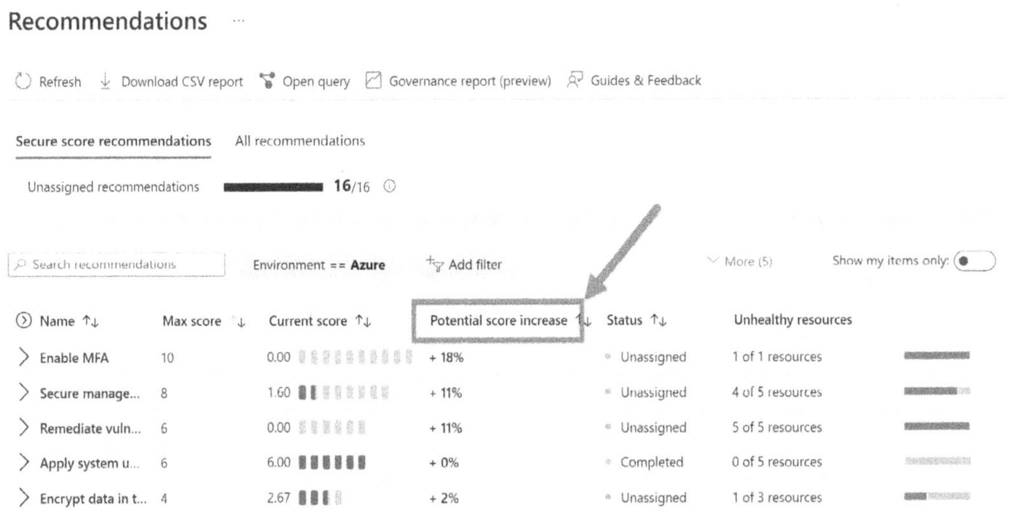

Recommendations ...

⟳ Refresh   ↓ Download CSV report   ⚡ Open query   ⌷ Governance report (preview)   ⅋ Guides & Feedback

Secure score recommendations   All recommendations

Unassigned recommendations   ████████ **16**/16 ⓘ

⊙ Name ↑↓	Max score ↑↓	Current score ↑↓	Potential score increase ↑↓	Status ↑↓	Unhealthy resources	
› Enable MFA	10	0.00 ▫▫▫▫▫▫▫▫▫▫	+ 18%	◦ Unassigned	1 of 1 resources	▬▬▬▬▬
› Secure manage...	8	1.60 ▮▮▫▫▫▫▫▫▫▫	+ 11%	◦ Unassigned	4 of 5 resources	▬▬▬▬▬
› Remediate vuln...	6	0.00 ▫▫▫▫▫▫	+ 11%	◦ Unassigned	5 of 5 resources	▬▬▬▬▬
› Apply system u...	6	6.00 ▮▮▮▮▮▮	+ 0%	◦ Completed	0 of 5 resources	▬▬▬▬▬
› Encrypt data in t...	4	2.67 ▮▮▮▫	+ 2%	◦ Unassigned	1 of 3 resources	▬▬▬▬

Environment == **Azure**   ＋ Add filter   ⌄ More (5)   Show my items only: ⦿

Figure 6.13: Potential score increase

Implementing the recommendations with the **Enable MFA** controls will increase the environment's secure score by 18% based on *Figure 6.13*. This is a highly impactful control and recommendation when evaluating this information. As a cybersecurity architect, you should evaluate the steps to remediate these controls by going deeper into the actions to take on resources by using Microsoft Defender for Cloud for CSPM.

Figure 6.14 shows the additional control recommendations for **Enable MFA**. Two are recommended:

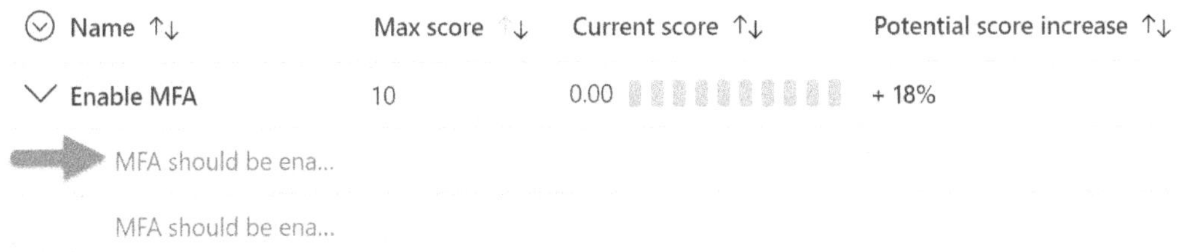

⊙ Name ↑↓	Max score ↑↓	Current score ↑↓	Potential score increase ↑↓
⌄ Enable MFA	10	0.00 ▫▫▫▫▫▫▫▫▫▫	+ 18%
MFA should be ena...			
MFA should be ena...			

Figure 6.14: Enable MFA recommendations list

Selecting the first in the list will take you to an overview of the recommendation, which includes the resources affected, along with a description and remediation steps. *Figure 6.15* shows the steps recommended to remediate and implement the security control to improve the secure score and overall security posture:

Home > Microsoft Defender for Cloud | Security posture > Recommendations >

MFA should be enabled on accounts with owner permissions on subscriptions   ⋯

⊘ Exempt   ◎ View policy definition   ⚓ Open query

ⓘ Multiple changes to identity recommendations will be available soon. Learn more →

∧ **Description**

Multi-Factor Authentication (MFA) should be enabled for all subscription accounts with owner permissions to prevent a breach of accounts or resources.

∧ **Remediation steps** ◁

Manual remediation:

To enable MFA using conditional access you must have an Azure AD Premium license and have AD tenant admin permissions.

1. Select the relevant subscription or click 'Take action' if it's available. The list of user accounts without MFA appears.
2. Click 'Continue'. The Azure AD Conditional Access page appears.
3. In the Conditional Access page, add the list of users to a policy (create a policy if one doesn't exist).
4. For your conditional access policy, ensure the following:
   a. In the 'Access controls' section, multi-factor authentication is granted.
   b. In the 'Cloud Apps or actions' section's 'Include' tab, check that Application Id for 'Microsoft Azure Management' App or 'All apps' is selected. In the 'Exclude' tab, check that it is not

Figure 6.15: Remediation steps within the security posture recommendations

The **Security posture** area and evaluating the secure score should be your first step in Microsoft Defender for Cloud for reviewing the security health of your Azure, hybrid, and multi-cloud environments.

> **Note**
> Microsoft 365 Defender has a similar dashboard for SaaS applications and workspace devices. This can be further reviewed and evaluated at `https://security.microsoft.com`. The format is very similar to Microsoft Defender for Cloud.

In the next section, you will learn more about CWPP within Microsoft Defender for Cloud through Defender plans.

Evaluating the Security Posture of Cloud Workloads

Microsoft Defender for Cloud provides enhanced **XDR** capabilities for workloads within Azure, AWS, Google, and on-premises/hybrid architectures with the various Defender plans for those workloads. These capabilities, along with Microsoft 365 Defender, provide a complete XDR solution for companies to secure their IaaS, PaaS, and SaaS workloads. Let's look at some resources that can be monitored and protected with Microsoft Defender for Cloud's enhanced security plans.

Defender for Servers provides threat protection for Azure virtual machines and non-Azure servers (including server **endpoint detection and response (EDR)**). Using hybrid infrastructure tools such as Azure Arc and Microsoft Defender for Endpoint, you can monitor and manage Azure and non-Azure servers. Non-Azure servers include those that are on-premises (or any physical or virtual server with an operating system), including AWS EC2 instances and GCP Compute Engine.

Microsoft Defender for Cloud secures AWS resources by implementing federated authentication. It starts with acquiring a Microsoft Entra token, which is then exchanged for AWS short-lived credentials. This is facilitated through an identity provider set up via OpenID Connect and IAM roles defined in a CloudFormation template. Defender for Cloud's CSPM service assumes the CSPM IAM role after the Microsoft Entra token is validated, ensuring secure access to AWS resources.

In the case of GCP, Microsoft Defender for Cloud uses a similar federated approach but tailored to Google Cloud's authentication mechanisms. It involves creating a workload identity pool and providers, along with service accounts and policy bindings, using the cloud template. The CSPM service acquires a Microsoft Entra token and exchanges it with Google's **Security Token Service (STS)** token. After validation, the Google STS token is returned to the CSPM service, allowing it to securely manage GCP resources.

Both processes are designed to maintain a high level of security and provide a unified security management experience across multi-cloud environments.

Additional Defender plans provide CWPP for supported Azure PaaS services. These additional plans, for an additional cost, also provide threat protection for the following:

- App services with Defender for App Service
- Containers (Azure containers and Azure Kubernetes Service) and container registries with Defender for Containers
- Azure DNS with Defender for DNS
- Azure Key Vault with Defender for Key Vault
- SQL, Cosmos DB, and open-source relational databases with Defender for Databases
- Azure storage accounts with Defender for Storage
- Resource management with Defender for Resource Manager

Figure 6.16 shows how the various workloads within Azure and non-Azure infrastructures can be monitored and managed when turning on enhanced security features in the Defender plans:

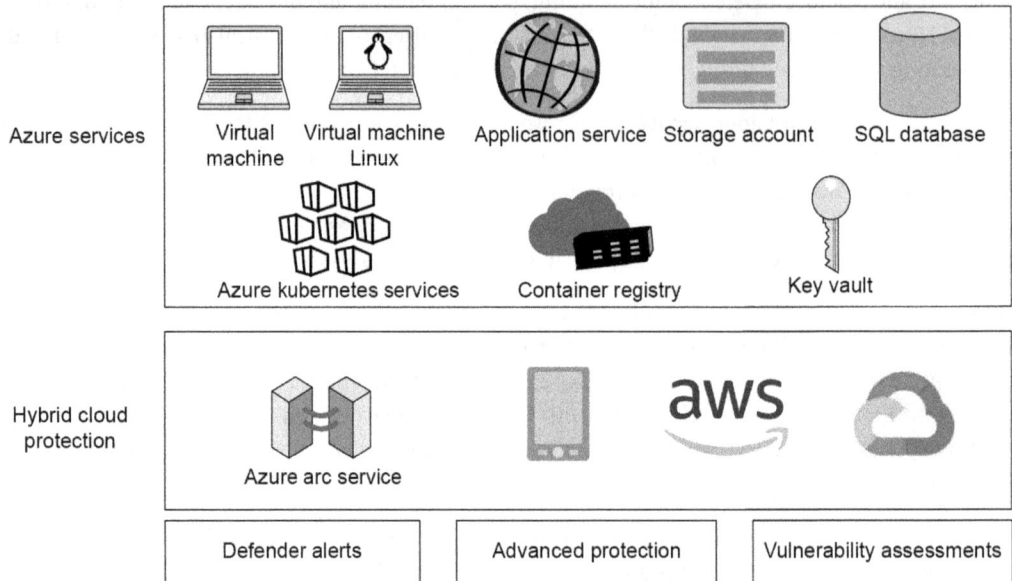

Figure 6.16: Microsoft Defender for Cloud enhanced security protection

More information on each of these plans can be found at this link: `https://docs.microsoft.com/en-us/azure/defender-for-cloud/defender-for-cloud-introduction`.

At the center of monitoring and managing an on-premises hybrid infrastructure is Azure Arc. Azure Arc bridges the ability to monitor hybrid workloads by gathering log data into Azure.

Now that you have understood the features and functionality of Microsoft Defender for Cloud, the next step is to configure your resources to ensure that Microsoft Defender for Cloud can provide appropriate recommendations for each of the resources that are connected:

1. Let's start by logging in to `https://portal.azure.com`. Then, in the search bar, enter `Microsoft Defender for Cloud`. Select **Microsoft Defender for Cloud**, as shown in *Figure 6.17*:

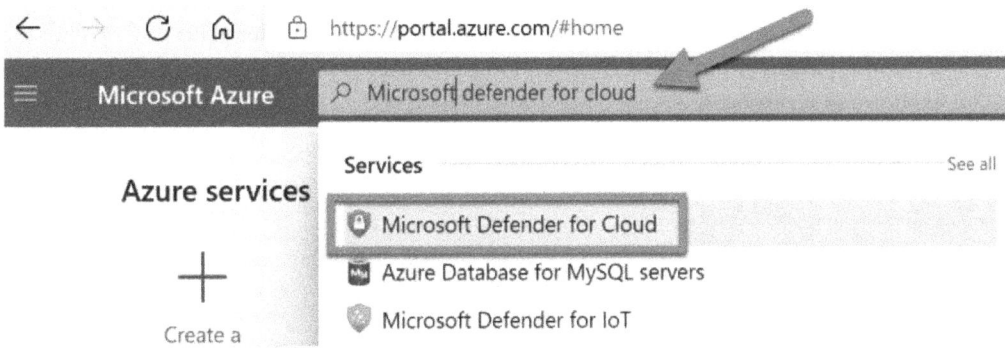

Figure 6.17: Navigating to Microsoft Defender for Cloud

This will take you to the Microsoft Defender for Cloud **Overview** tile.

2. In the **Overview** tile, locate **Environment settings** in the **Management** menu, as shown in *Figure 6.18*:

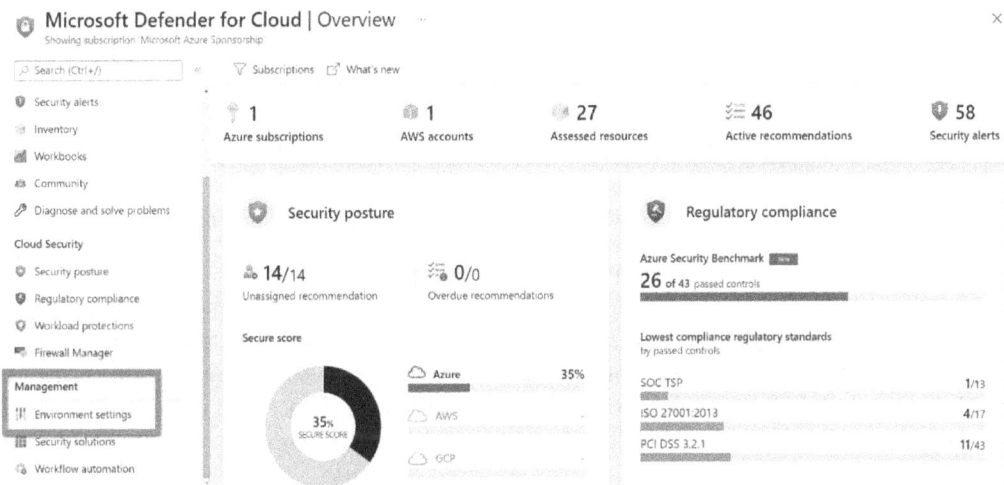

Figure 6.18: Environment settings in Microsoft Defender for Cloud

3. Select your subscription. This will take you to the **Settings | Defender plans** page. Within the **Defender plans** settings, select **Enable all** to enable all Microsoft Defender for Cloud plans, as shown in *Figure 6.19*:

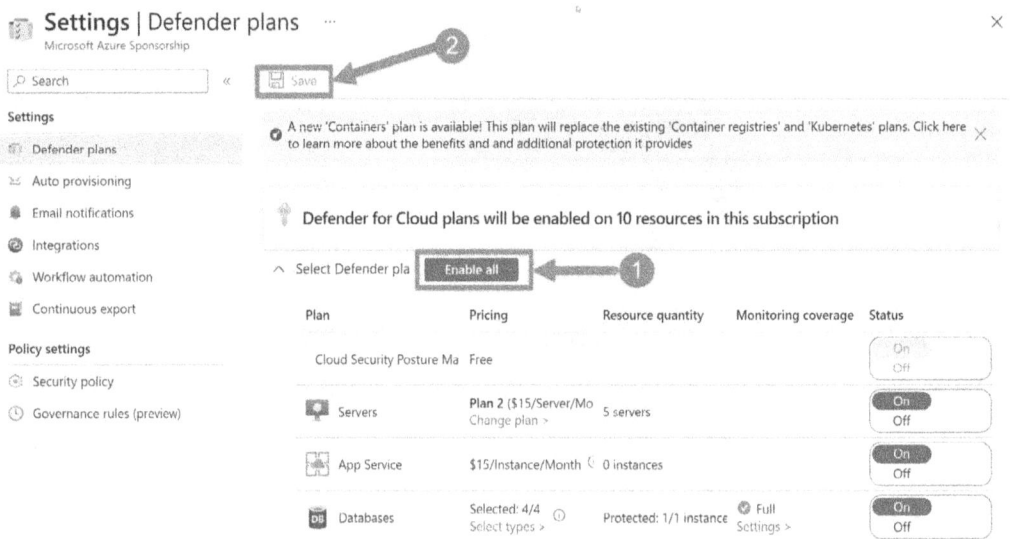

Figure 6.19: Enabling Defender for Cloud enhanced security plans

This will allow you to manage the security posture for hybrid and multi-cloud resources.

When enabling the Defender for Cloud plans, you enable the additional features available for Microsoft Defender for Cloud. Currently, CSPM features are free for customers. Among the additional features enabled with Defender, plans include **Just in Time Virtual Machine Access**, adaptive application controls, network hardening, threat protection for servers, and PaaS. When these Defender plans are enabled, you unlock additional vulnerability and threat alerts that can be accessed within the **Workload protections** section under the **Cloud Security** menu, as shown in *Figure 6.20*:

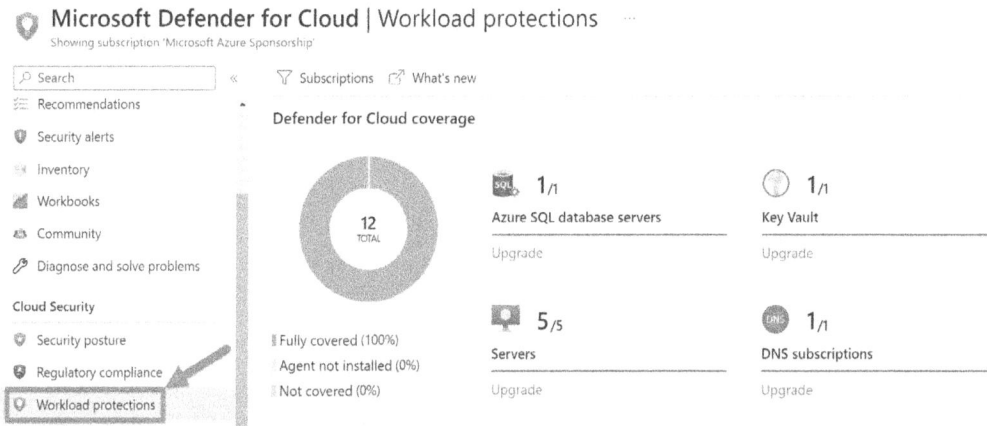

Figure 6.20: Workload protections in the Cloud Security menu

Here is an example of a solution and a problem. **Contoso** is a financial services company that stores vast amounts of sensitive customer data across multiple cloud platforms, including AWS and GCP. Despite having a robust IT team, it faced a significant challenge when confidential client information was inadvertently exposed due to a misconfigured network security group. This breach not only led to financial penalties but also damaged its reputation.

The root cause of the breach was identified as a lack of centralized security management and real-time threat detection across its multi-cloud environment. Contoso's security team struggled to keep up with the evolving security configurations and compliance requirements necessary to protect its cloud-based assets.

By leveraging Microsoft Defender for Cloud, Contoso not only resolved its immediate security issues but also established a proactive defense mechanism that continuously adapts to new threats, ensuring the ongoing protection of its cloud resources and customer data. This strategic move restored its clients' trust and positioned Contoso as a leader in data security within the financial industry.

Within the **Workload protections** dashboard, you can scroll down and review the security posture for the workloads within the scope of the enabled Defender plans, as shown in *Figure 6.21*:

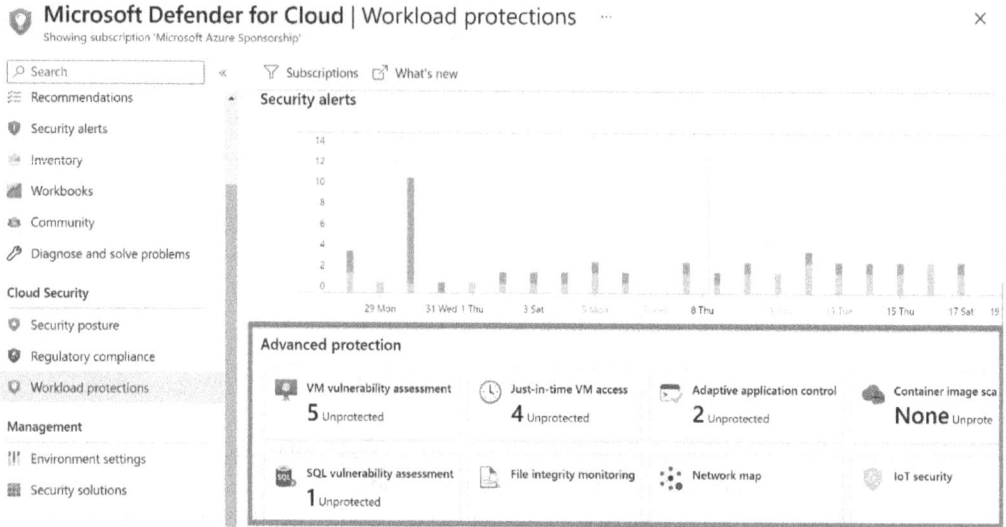

Figure 6.21: Advanced protection for Defender plans

Selecting one of these workloads will provide additional recommendations for hardening the security posture of these workloads. The **Network Map** area is a very helpful feature that you can use to evaluate potential exposure within virtual networks and virtual machines. *Figure 6.22* shows how you can use this network map to view additional recommendations for network and virtual machine hardening. The red exclamation points denote potential vulnerabilities and areas of recommended security improvements:

Network Map

Showing subscription 'Microsoft Azure Sponsorship'

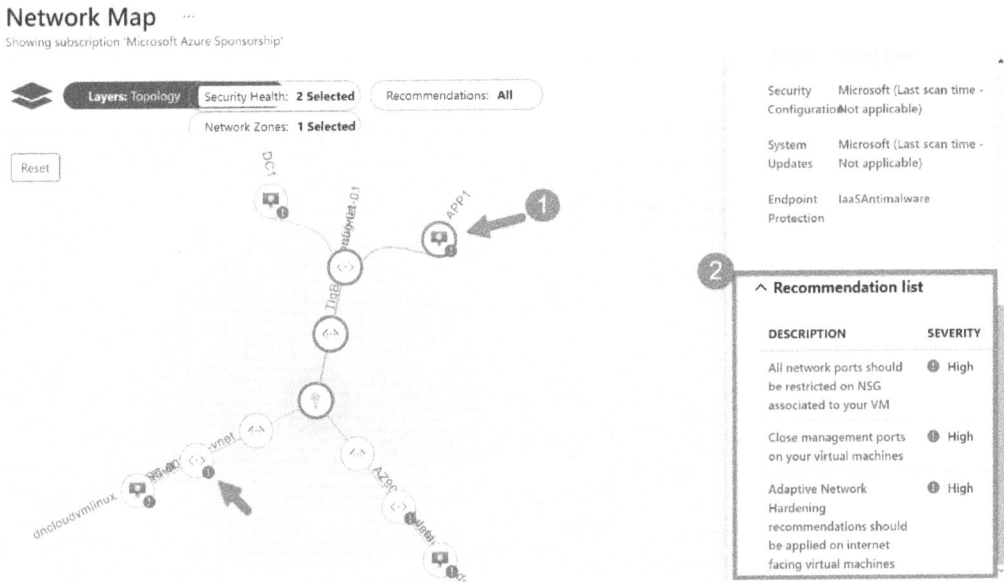

Figure 6.22: Network map and security posture recommendations

Workload protections within Microsoft Defender for Cloud provide helpful insights into the security posture and hygiene of the infrastructure and overall environment. These insights and alerts allow you to address potential issues and put controls in place before an attack.

The next section will discuss how to design security for an Azure landing zone.

Designing Security for an Azure Landing Zone

The cloud landing zone should be treated as the foundation for a cloud infrastructure deployment. The landing zone encapsulates the identity and access management roles, policies that govern the environment, the monitoring and management connections, the application and compute parameters and templates, and the overall security and governance for resources that are deployed within the subscription.

The design of your Azure landing zone should have a strong focus on security within each of the key areas deployed within the landing zone. Azure landing zones are deployed by templates and code. Companies generally refer to this with the words "secure by design." To ensure this is a reality and not just a shallow statement, cybersecurity architects should be involved in the design of these landing zones with the cloud architects. *Figure 6.23* shows the various components of the landing zone:

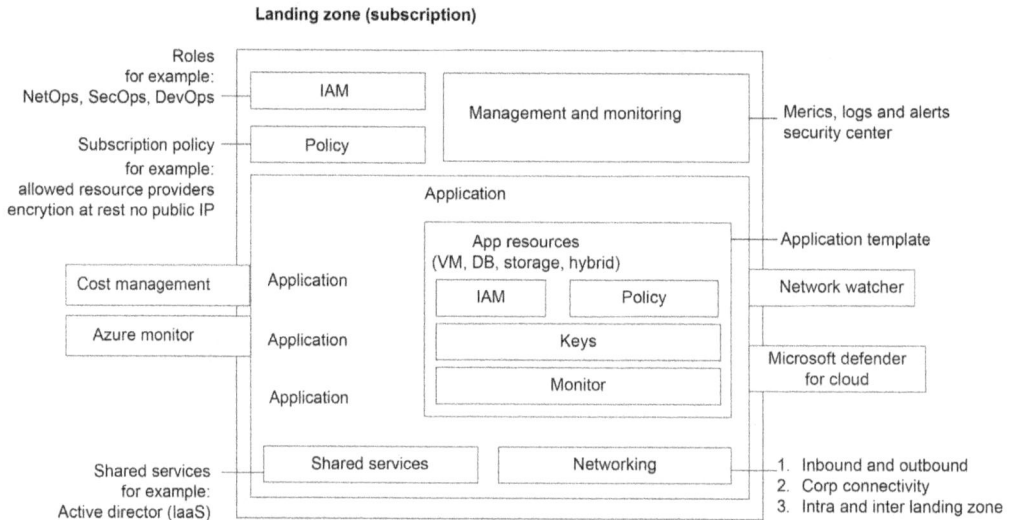

Figure 6.23: Diagram of the components of an Azure landing zone

Each of the components of an Azure landing zone should have some aspect of security and governance for identities, apps and infrastructure, networking, key and secret management, and overall monitoring and management.

The following table provides the areas of focus for secure design:

Design Area and Methodology	Objective	Microsoft Documentation
Security	Provides controls and protection processes for securing cloud environments	`https://docs.microsoft.com/ en-us/azure/cloud-adoption- framework/ready/landing-zone/ design-area/security`
Management	Creates ongoing operations procedures and management baselines, as well as protection and recovery capabilities	`https://docs.microsoft.com/ en-us/azure/cloud-adoption- framework/ready/landing-zone/ design-area/management`

Design Area and Methodology	Objective	Microsoft Documentation
Governance	Provides policies to automate auditing and enforcement of compliance	`https://docs.microsoft.com/ en-us/azure/cloud-adoption- framework/ready/landing-zone/ design-area/governance`
Ready and automation	Utilizes tools and templates to deploy and automate the creation of landing zones and resources	`https://docs.microsoft.com/ en-us/azure/cloud-adoption- framework/ready/landing- zone/design-area/platform- automation-devops`

Table 6.1: Azure landing zone design areas and methodology

When designing and determining the security controls for an Azure landing zone, the cybersecurity architect should review and determine that the following design considerations are part of the deployment:

- Identity and access controls
- Security alerts
- Security logs
- Security controls
- Vulnerability management
- Shared responsibility for controls
- Encryption and keys
- Backup and recovery

Integrating the tools provided within Azure with Microsoft Defender for Cloud, Microsoft Sentinel, Azure DDoS standard protection, Privileged Identity Management, and firewall protection with Azure Firewall and **Web Application Firewall** (**WAF**) helps with creating the security controls for the Azure landing zone and the overall environment.

The Azure landing zone provides the deployments within the Azure subscription with the building blocks that become the foundation for the security and governance of your resources. In the next section, we will discuss how to interpret threat intelligence and provide recommendations for mitigating risk.

Interpreting Technical Threat Intelligence and Recommending Risk Mitigations

Microsoft is at the forefront globally for reviewing and recognizing threats through its alliances and participation in the **cyber threat intelligence** (**CTI**) network. The information that is gathered through the CTI reports, communities, investigation feeds, and organizational security investigations are used within Microsoft's cloud services for customers to identify threats and vulnerabilities within their environments.

SIEM solutions are the primary tools that customers can use for evaluating CTI. Within Microsoft and Azure, that solution is Microsoft Sentinel. Microsoft Sentinel utilizes CTI from a variety of security solutions within Microsoft and other third-party solutions to provide a single source to identify and interpret potential threats and attacks within your company environment.

Figure 6.24 shows this flow of information within Microsoft Sentinel:

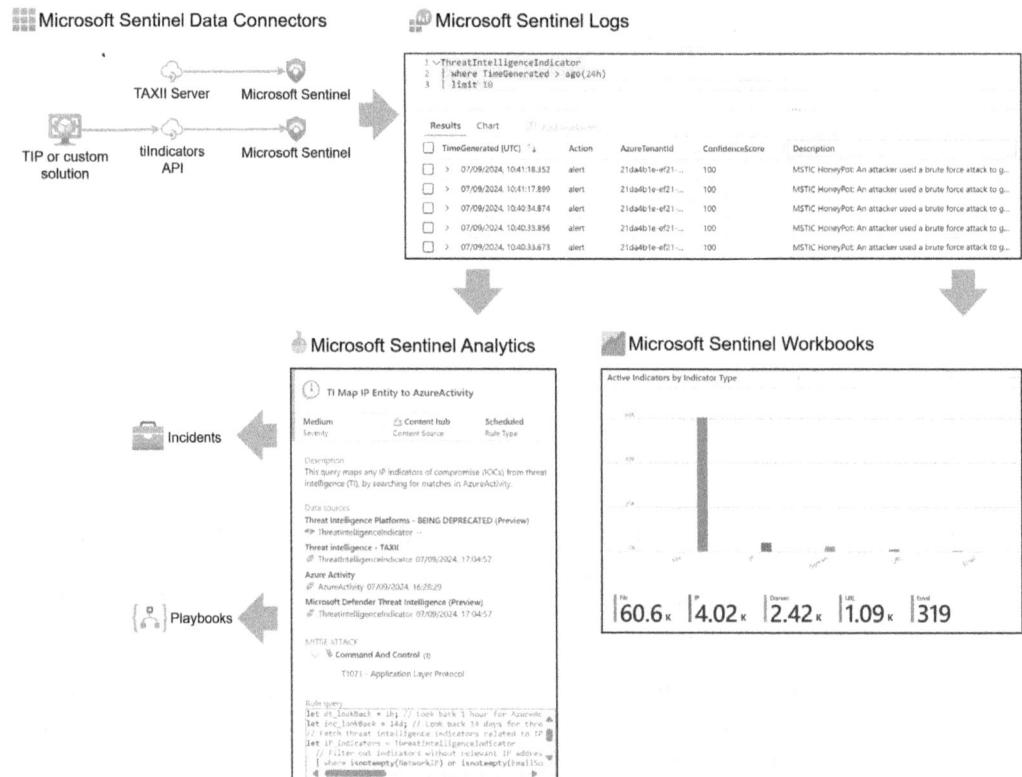

Figure 6.24: Microsoft Sentinel tools for threat intelligence

The log queries, workbooks, and analytics rules within Microsoft Sentinel provide information that can be used to investigate potential threats and vulnerabilities. This information can also be used to hunt for possible attacks or pre-attacks that are taking place within your cloud environment.

In addition to Microsoft Sentinel, Microsoft Defender for Cloud can also be used as a solution for CTI. The alerts and vulnerability assessments within the **Workload protection** dashboard can be used to investigate and identify potential vulnerabilities and possible pre-attacks that are happening within your environment.

Part of managing and interpreting threat intelligence and the overall security posture of your organization is identifying and mitigating risk. Threat and risk analysis is a process that every company should go through to properly understand the vulnerabilities within their systems and the potential threat to the company if that vulnerability is exploited.

When doing a threat and risk analysis, you need to consider the perceived risk to an asset. To accomplish this, you must consider the combination of the asset and any vulnerabilities, along with the potential threat that the vulnerability will be exploited. As an equation, this would look like this:

Asset (A) + Vulnerability (V) + Threat (T) = Risk (R)

The assessment of the risks to assets can be categorized into various levels of risk, as shown in *Figure 6.25*:

Likelihood		Minor	Moderate	Major
	Very likely	Acceptable risk medium 2	Unacceptable risk high 3	Unacceptable risk extreme 5
	Likely	Acceptable risk low 1	Acceptable risk medium 2	Unacceptable risk high 3
	Unlikely	Acceptable risk low 1	Acceptable risk low 1	Acceptable risk medium 2
	What is the chance that it will happen?	Minor	Moderate	Major

Impact
How Serious is the Risk?

Figure 6.25: Risk assessment matrix

Identifying the company assets and then assessing the risk should be done at a business and technical level to have a proper understanding and perception across departments.

Once all of the assets and levels of risk have been identified, then the analysis of financial exposure needs to be determined. This is accomplished by determining the following criteria for each asset:

- **Exposure factor** (**EF**) is the impact that is measured by the percentage of loss of an asset if the risk is realized.

- **Single-loss expectancy** (**SLE**) is the value of the asset multiplied by the EF. This is going to place a financial value on the asset loss when exposed.

- **Annualized rate of occurrence** (**ARO**) places the possible number of times that this risk may be exploited over the year.

- **Annualized loss expectancy** (**ALE**) is the combined financial impact of the SLE times the ARO, providing an annual cost of loss to the asset.

Calculating the ALE for all of the company's digital assets places a tangible value on those assets, which then allows a company to evaluate the cost of investing in the controls needed to mitigate or avoid that risk. If a company is not going through these steps, they are not placing proper due care and due diligence on protecting company assets.

Another point to make here is that this is placing a financial value on assets and not reputational value. The reputational impact of data exposure or data loss can be much more damaging to a company with its customers than the perceived value of the assets. Companies should consider this when determining investment in security controls as well.

As stated in the previous section, data residency is an important aspect of protecting the privacy of data, both business and personal.

Recommending Security Capabilities or Controls to Mitigate Identified Risks

Once you have evaluated and identified the risks, you determine how to mitigate them. The life cycle of risk analysis is a continuous cycle of identification, assessment, response, monitoring, and reporting. *Figure 6.26* shows this life cycle and the continuous cycle of identifying and mitigating risks:

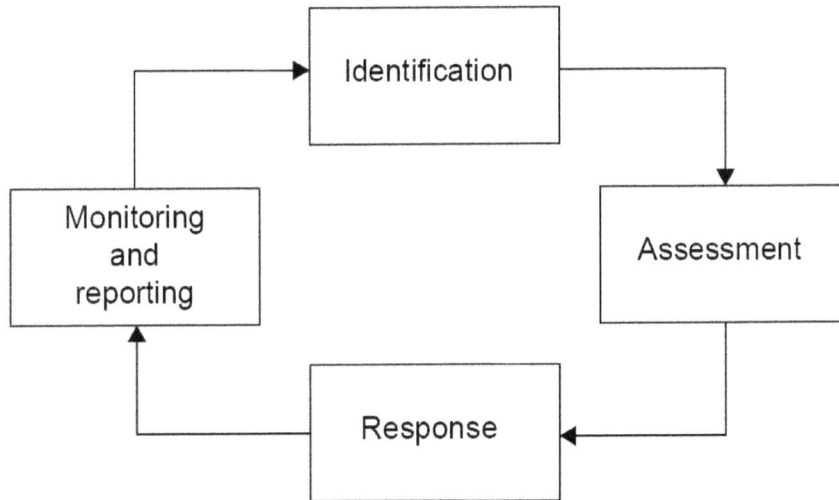

Figure 6.26: Risk assessment and mitigation life cycle

Throughout this chapter, you have learned about tools within Microsoft Defender for Cloud that help you identify potential risks and vulnerabilities that could create threats and attacks on your company's environment. Whether you are using regulatory compliance, security posture, and Secure Score or workload protection, you are reviewing and analyzing risks and gathering recommendations for controls that you can put in place to mitigate these risks. As a cybersecurity architect, you should utilize these tools and recommend their use within the cloud architecture and design. When you are brought in to analyze and determine risks, these tools can identify potential problem areas where you may need further controls to be put in place.

In the next section, you will be given a company scenario and asked to complete several tasks to meet the requirements to determine potential vulnerabilities and risks to the environment and recommend controls.

Evaluating the Security of Internet Assets with Microsoft Defender EASM

Microsoft Defender EASM provides a holistic view of your security posture by continuously discovering and mapping your digital attack surface. Implement continuous inventory monitoring to detect, analyze, and classify external-facing resources as they emerge, thereby extending your inventory control beyond the firewall. Focus on identifying and addressing vulnerabilities and misconfigurations, and bring these resources under management to reduce risks. Additionally, use generative AI insights from Microsoft Copilot for Security to swiftly pinpoint risky assets and maintain a comprehensive view of your external risk posture.

EASM is a growing trend in cybersecurity that focuses on identifying and managing the external-facing assets of an organization. It provides an objective, outside-in view of your digital footprint, which is crucial for maintaining a robust security posture. Think of it as walking the edges of a castle, scanning for weaknesses that attackers might exploit to gain entry.

By utilizing EASM, organizations can continuously monitor and analyze their external-facing resources, such as websites, IP addresses, and cloud services, to detect vulnerabilities and misconfigurations. This automated approach extends your inventory control beyond the firewall, ensuring that all potential entry points are identified and secured.

The importance of EASM lies in its ability to prioritize vulnerabilities and bring these resources under management, thereby mitigating risks. Additionally, leveraging generative AI insights from Microsoft Copilot for Security allows for the quick identification of risky assets, helping to maintain a unified view of your external risk posture.

Let's dive deeper into designing a solution for evaluating the security posture of internet-facing assets using Defender EASM.

Dynamic Inventory Monitoring

This feature continuously scans your external attack surface, identifying new assets as they appear. Provided your inventory is up to date, you can use this feature to quickly spot any unexpected or unauthorized resources. By continuously monitoring and analyzing external-facing resources, such as websites, IP addresses, and cloud services, EASM helps detect vulnerabilities and misconfigurations, extending inventory control beyond the firewall. EASM is particularly important for cloud environments, where it evaluates the security posture of cloud workloads, including virtual machines, containers, and serverless functions. It also extends to APIs, ensuring they are secure and preventing unauthorized access. Additionally, EASM encompasses third-party services, SaaS applications, and IoT devices, ensuring all external-facing assets are monitored and managed effectively.

Multi-Cloud Visibility

Organizations often operate across multiple cloud providers and hybrid environments. Defender EASM ensures that your inventory control isn't limited to a single cloud. It tracks assets across Azure, AWS, Google Cloud, and other platforms. This comprehensive view helps you understand the full scope of your attack surface, including third-party services and shadow IT.

Risk Prioritization

Not all assets are equally critical. Some may host sensitive data, while others are less impactful. Defender EASM assesses risk by prioritizing vulnerabilities and misconfigurations. It considers factors such as asset importance, exposure level, and potential impact. By focusing on high-risk items first, you can allocate resources effectively and address the most urgent security gaps.

Generative AI Insights

Microsoft Copilot for Security leverages AI to provide actionable insights. Copilot analyzes data from Defender EASM to identify patterns, anomalies, and emerging threats. It's like having an intelligent security advisor – one that highlights risky assets, recommends mitigation steps, and keeps you informed about the evolving threat landscape.

Figure 6.27 shows the Microsoft Defender EASM workspace right after it is created in the Azure portal.

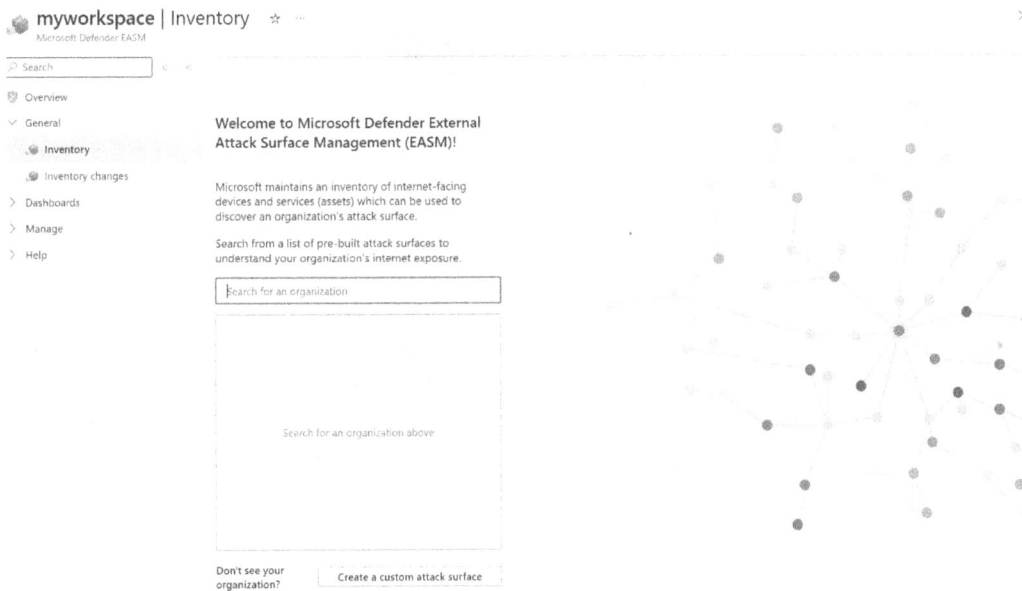

Figure 6.27: Screenshot of Defender for EASM in Azure portal

Remember, this integrated approach, combining Defender EASM's continuous monitoring with Copilot's AI-driven insights, strengthens your organization's security posture. By proactively managing external risks, you're better equipped to defend against cyber threats.

Case Study – Evaluating the Security Posture

Apply what you learned in this chapter by completing the case study on the accompanying online platform. In this case study, you will be given a company scenario and asked to complete several tasks to ensure the company meets the standards by evaluating its security posture.

To access the case study, visit the following link or scan the QR code.

Link to the case study: `https://packt.link/SC100-E2-CaseStudy_Chapter6`

QR code:

Figure 6.28: QR code to access case study for Chapter 6

Summary

In this chapter, we discussed how to evaluate the security posture of a company for its Azure, hybrid, and multi-cloud environments by utilizing the tools within Microsoft Defender for Cloud. This included evaluating security controls for regulatory compliance, including using the Azure Security Benchmark, Secure Score recommendations, and workload protections to alert and protect against vulnerabilities and threats on compute, storage, and networks. You learned how to design security in Azure landing zones and evaluate threat intelligence and risks to determine mitigation.

In the next chapter, you will learn how to design a security strategy for on-premises and cloud virtual machines, as well as other networked endpoints.

Exam Readiness Drill – Chapter Review Section

Apart from mastering key concepts, strong test-taking skills under time pressure are essential for acing your certification exam. That's why developing these abilities early in your learning journey is critical.

Exam readiness drills, using the free online practice resources provided with this book, help you progressively improve your time management and test-taking skills while reinforcing the key concepts you've learned.

How to Get Started

1. Open the link or scan the QR code at the bottom of this page.

2. If you have unlocked the practice resources already, log in to your registered account. If you haven't, follow the instructions in *Chapter 11* and come back to this page.

3. Once you have logged in, click the **START** button to start a quiz.

We recommend attempting a quiz multiple times till you're able to answer most of the questions correctly and well within the time limit.

You can use the following practice template to help you plan your attempts:

Working On Accuracy		
Attempt	Target	Time Limit
Attempt 1	40% or more	Till the timer runs out
Attempt 2	60% or more	Till the timer runs out
Attempt 3	75% or more	Till the timer runs out
Working On Timing		
Attempt 4	75% or more	1 minute before time limit
Attempt 5	75% or more	2 minutes before time limit
Attempt 6	75% or more	3 minutes before time limit

The above drill is just an example. Design your drills based on your own goals and make the most of the online quizzes accompanying this book.

First time accessing the online resources? 🔒

You'll need to unlock them through a one-time process. **Head to** *Chapter 11* **for instructions.**

Open Quiz	
https://packt.link/SC100_CH06	
Or scan this QR code →	

7

Design a Strategy for Securing Server and Client Endpoints

In *Chapter 6*, you explored how to evaluate the security posture using the tools provided in Microsoft Defender for Cloud and recommended strategies to manage risk and vulnerabilities. Building on that foundation, this chapter will focus on designing a comprehensive strategy for securing servers and client endpoints.

You will delve into creating a security baseline for both server and client endpoints, specifying security requirements for servers, mobile devices, and **Active Directory Domain Services (AD DS)**. Additionally, you will learn how to design strategies for managing secrets, keys, and certificates, as well as ensuring secure remote access for endpoints.

This chapter aligns with the exam domains and tasks for **designing security operations, identity, and compliance capabilities**, as well as for **designing security solutions for infrastructure**.

In this chapter, you are going to cover the following main topics:

- Planning and implementing a security strategy across teams
- Specifying security baselines for server and client endpoints
- Specifying security requirements for servers, including multiple platforms and operating systems
- Specifying security requirements for mobile devices and clients, including endpoint protection, hardening, and configuration
- Specifying security requirements for IoT devices and connected systems
- Evaluating solutions for securing **operational technology** (OT) and **industrial control systems** (ICSs) by using Microsoft Defender for IoT
- Evaluating Windows **Local Admin Password Solution (LAPS)** solutions
- Specifying requirements to secure AD DS
- Designing a strategy to manage secrets, keys, and certificates

- Designing a strategy for secure remote access
- Understanding security operations frameworks, processes, and procedures
- Case study – designing a secure architecture for endpoints

Planning and Implementing a Security Strategy across Teams

You should have a strategy for security across all stakeholders. This strategy is a key component of managing the security of your infrastructure. Security teams, networking teams, and infrastructure administration teams must all work together and have clearly defined roles to manage and maintain security baselines and continued protection for infrastructure, endpoints, identities, and data.

Cloud security requires a different way of thinking than traditional on-premises data center security. The physical security of the building, data center, and network interfaces is no longer in house, and the IT and security professionals must collaborate on a consistent strategy to implement, monitor, and manage infrastructure, devices, identity, and data. This requires a plan that includes defining processes and guidelines that will make up the security baselines for the company. It will also need to incorporate training plans to ensure that all stakeholders understand cybersecurity in the cloud and their roles/responsibilities within it.

Figure 7.1 outlines the full scope of roles and responsibilities across the full process of planning, building, and running the environment and the different security touch points for leadership, architects, and administrators:

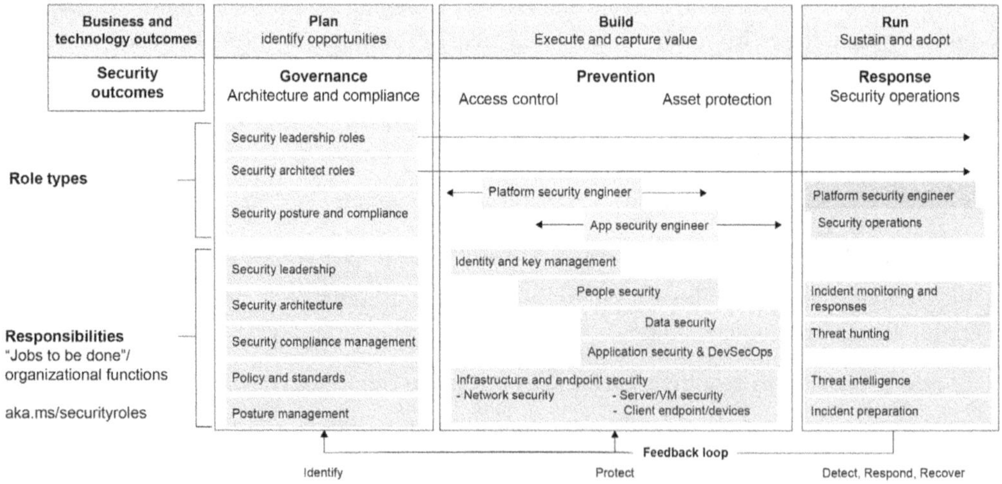

Figure 7.1: A flowchart that outlines a structured approach to cybersecurity, divided into five phases: Govern, Plan, Build, Operate, and Sustain

The key point to take away from this diagram is that a security architect's involvement is not just in planning and building the infrastructure – it must be continually involved throughout. Security requirements are constantly changing and need involvement from the entire team.

> **Note**
>
> Additional details on these roles and responsibilities can be found at `https://learn.microsoft.com/en-us/azure/cloud-adoption-framework/organize/cloud-security`.

This continuous assessment and validation process is also captured within the Microsoft **Cloud Adoption Framework (CAF)** for Azure.

The Microsoft CAF is a comprehensive guide designed to help you navigate the complexities of cloud adoption. It provides proven guidance, best practices, and tools to support your organization in creating and implementing effective business and technology strategies for the cloud. By leveraging the CAF, you can confidently align your cloud adoption efforts with your business objectives, ensuring a smooth transition and maximizing the benefits of cloud technologies.

As you embark on your cloud journey, the CAF will assist you in defining your strategy, planning your approach, and preparing your organization for the cloud. It covers every stage of the cloud adoption life cycle, from initial planning and readiness to ongoing governance and management. With the CAF, you can achieve operational excellence, secure your environment, and drive innovation, all while managing risks and optimizing costs.

> **Note**
>
> Security guidelines for the CAF can be found here: `https://learn.microsoft.com/en-us/azure/cloud-adoption-framework/secure/`.

Figure 7.2 shows the CAF process for cloud adoption, with continuous security, management, and governance being cyclical:

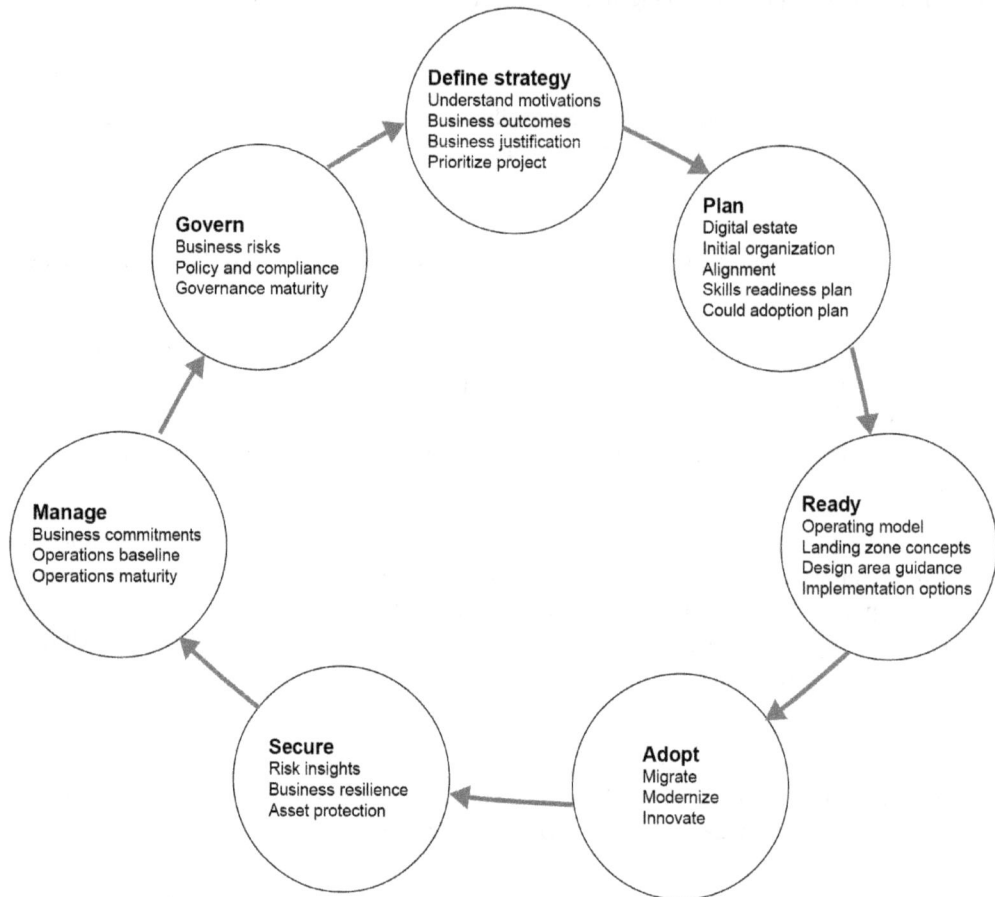

Figure 7.2: CAF depicted as a lifecycle: define strategy, plan, ready, adopt, secure, manage, govern, and then start again

The bottom part of the CAF outlines the shift-left approach to continuously improving security posture and reviewing the environment for risks and vulnerabilities. Meticulously planning the team roles and responsibilities and utilizing the CAF for quick wins along the cloud security journey will build confidence and create a solid foundation going forward.

The capabilities that are enabled for security posture management and Secure Score within Microsoft Defender for Cloud provide the tools that offer visibility into misconfigurations and improvement recommendations. These were discussed in *Chapter 6, Evaluate the Security Posture and Recommending Technical Strategies to Manage Risk*. The security management strategy should include the following stages:

- Continuously assess the infrastructure for the current security posture
- Secure the infrastructure by hardening endpoints, resources, and services
- Defend the infrastructure with security alerts to defend and resolve threats and vulnerabilities

The next section will discuss tools that you can use to specify security baselines for server and client endpoints.

Specifying Security Baselines for Server and Client Endpoints

The various endpoint devices are important when designing and building a strategy for zero-trust and defense in depth. Users utilize endpoint devices to access resources and server endpoints to make up critical infrastructure for applications. These devices are key to the company's ability to function but are also an attack surface where vulnerabilities can be exploited if due care and due diligence are not taken seriously.

That due care and due diligence require recognizing and understanding where the vulnerabilities are within server and client endpoints. Creating a proper security baseline and strategies for how users, groups, and teams utilize and access endpoints will reduce the attack surface on these devices.

Microsoft attempts to design its Windows client and server operating systems to be secure out of the box. However, the secure nature of the operating systems does not last as bad actors attempt to find vulnerabilities to exploit. As security professionals and administrators, it is our responsibility to have processes that define the security baseline that our company will adhere to for servers and client endpoints.

The following are some baseline principles for securing endpoints:

- For security-conscious companies, end users should not have administrative rights to their company-managed devices.
- The baseline security control should not create operational issues worse than the threat you are attempting to mitigate. In other words, you should not make it harder for the end user to use that endpoint.
- Baselines enforce security defaults to prevent an authorized user from selecting an insecure state for that endpoint.

Microsoft provides guidance and preconfigured settings within the server and client operating systems to build that baseline. This guidance includes downloading the **Security Compliance Toolkit (SCT)**.

The SCT is an essential resource for administrators aiming to manage security baselines effectively. It provides a suite of tools that allow you to download, analyze, test, edit, and store Microsoft-recommended security configuration baselines for Windows and other Microsoft products.

With the SCT, you can compare your current **Group Policy Objects** (**GPOs**) with Microsoft-recommended baselines or other security configurations. This comparison helps identify discrepancies and ensures that your security settings align with best practices. The toolkit includes tools such as the Policy Analyzer, which highlights redundant settings or internal inconsistencies within GPOs, and the **Local Group Policy Object (LGPO)** tool, which aids in automating the management of local policies.

By using the SCT, you can streamline the process of applying security baselines across your organization, ensuring a consistent and robust security posture.

> **Note**
>
> For more information on this guidance, you can review this link: `https://learn.microsoft.com/en-us/windows/security/operating-system-security/device-management/windows-security-configuration-framework/windows-security-baselines`.

Endpoint security within Microsoft Intune provides additional information and a dashboard where you manage these security baselines. *Figure 7.3* shows the operating systems and browsers that are supported for Windows-managed devices:

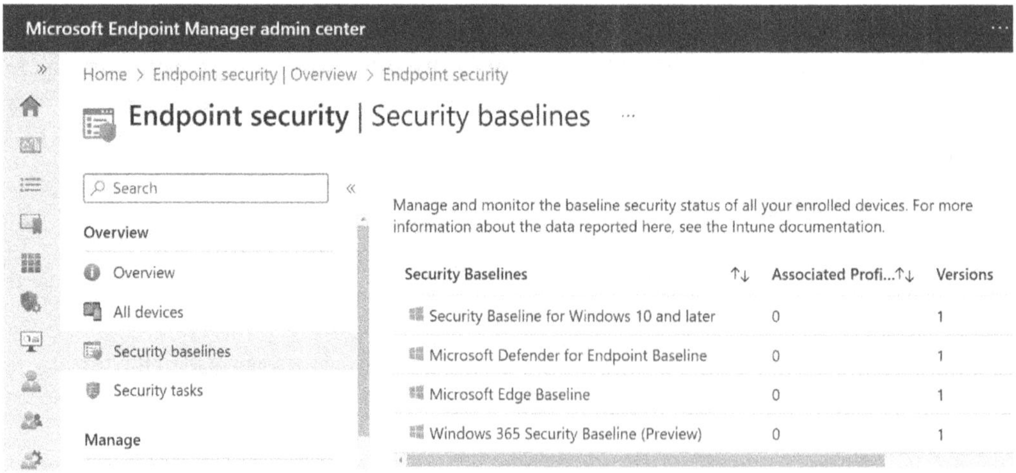

Figure 7.3: Microsoft Endpoint Manager admin center on the Endpoint security | Security baselines page

The security baselines for servers and client endpoints can be enforced through mobile device management and GPOs within Endpoint Manager. Microsoft Defender for Cloud can also be used to identify virtual machine vulnerabilities and recommend hardening the server operating system.

The SCT should be the starting point of your security baseline strategy for server and client endpoints. In the next section, you will learn how to specify server security requirements for different platforms and operating systems.

Specifying Security Requirements for Servers, Including Multiple Platforms and Operating Systems

Utilizing the SCT is one way to build a strategy for your server endpoints to maintain a security baseline. Included within the security strategy is how to manage the administration of the server operating systems. Each server and endpoint within a domain has an administrator account. In a large company, managing these administrator accounts becomes a challenge. These administrator accounts become a security concern and increase the attack surface for a potential identity breach to gain access to the account. The danger is that that server administrator account could be the administrator account to the domain controller on the AD DS network, exposing the domain controller and the member servers.

This means that if a server administrator account is compromised, it could potentially give an attacker access to the domain controller, which is the central authority in an Active Directory network. The domain controller manages all security and permissions for the network, including user accounts and access to resources.

The encrypted **Server Message Block (SMB)** protocol can be used to protect against exposure on these servers. The SMB protocol is primarily used on domains for file sharing on the network. Printers, scanners, and email servers also commonly use the SMB protocol. Enabling encryption for the SMB protocol will protect data exposure on storage accounts for file servers, email servers, and even SQL servers on the AD DS domain.

For hybrid infrastructures with Azure and on-premises server resources, a strategy that utilizes Azure Arc to communicate with Azure and Microsoft Defender for Cloud should be used. Utilizing Microsoft Defender for Cloud allows continuous assessment of the security posture for on-premises and hybrid resources against the Azure Security Benchmark. The Azure Security Benchmark will evaluate servers that are reporting through Azure Arc and provide recommendations for hardening those resources into the Microsoft Defender for Cloud security posture and workload protection. This will reduce the overall attack surface across server endpoints.

Figure 7.4 provides a diagram of this hybrid architecture:

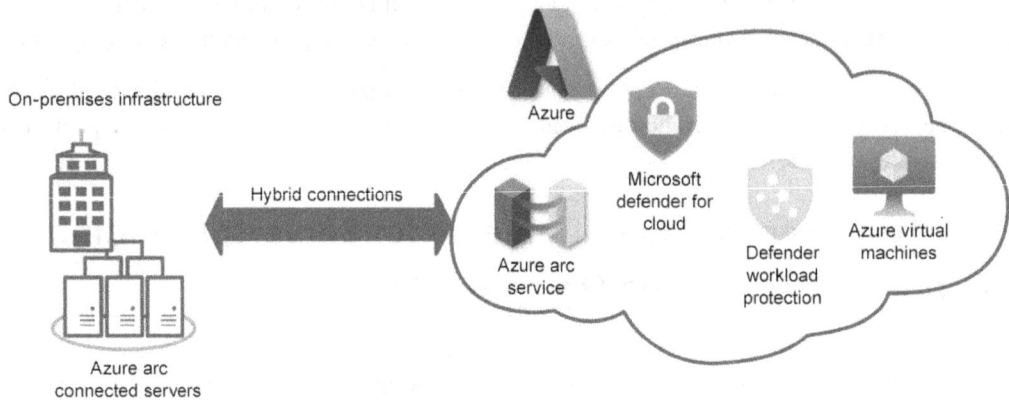

Figure 7.4: Hybrid server infrastructure

As *Figure 7.4* shows, servers connected on-premises communicate through Azure Arc into Azure. These servers, both Windows and Linux, become part of the virtual machine infrastructure that is evaluated for security vulnerabilities and threats within workload protection within Microsoft Defender for Cloud. The recommendations that are provided will harden these servers against the Azure Security Benchmark.

The next section will discuss how to determine and create a strategy for security for mobile devices.

Specifying Security Requirements for Mobile Devices and Clients, Including Endpoint Protection, Hardening, and Configuration

Mobile devices create a significant challenge to security personnel. The cybersecurity architect should help build a strategy that covers capabilities for hardening devices. The SCT can be used for security baselines on Windows devices, servers, and client endpoints. Mobile devices within a company are not simply Windows operating systems. Operating systems for these mobile devices also include iOS, macOS, and Android. In addition to the multiple operating systems, these devices are accessing resources from different networks, some of which are public and unsecured.

Security for these mobile devices requires a company to review and monitor that these devices are maintaining compliance with the company's security benchmarks. **Mobile device management (MDM)** or **mobile application management (MAM)** will allow device administrators and security administrators to have visibility of the compliance and protection of these devices, across multiple operating systems. At a minimum, company-owned devices should be managed with MDM. MDM provides full control of each operating system's configuration and compliance profiles. MAM will maintain compliance for company-maintained applications and a separation on the devices between personal and company data. MDM should be used, when possible, to manage all devices accessing company data and applications.

MDM and MAM are provided through Microsoft Intune, which is part of Microsoft Endpoint Manager within the Microsoft 365 suite of security solutions.

> **Note**
>
> More information on Microsoft Endpoint Manager can be found at this link: `https://learn.microsoft.com/en-us/mem/endpoint-manager-overview`.

When managing mobile devices with MDM and MAM, there is a life cycle in onboarding, managing, and removing these devices to prevent an increased attack surface. *Figure 7.5* shows this life cycle:

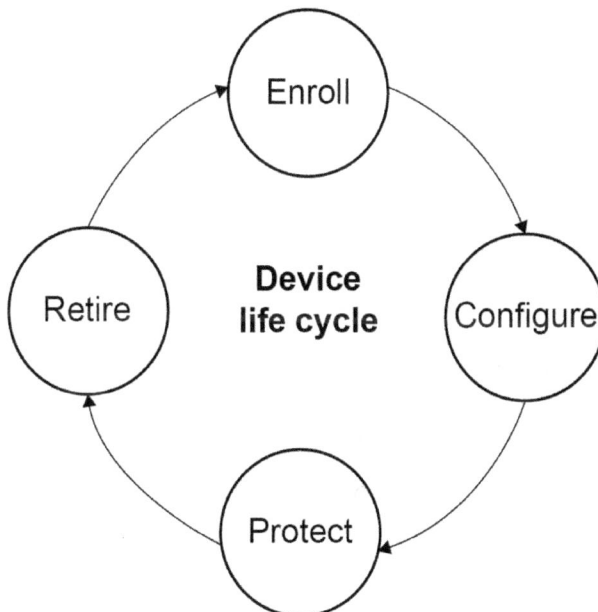

Figure 7.5: Mobile device life cycle: from Enroll, through Configure, Protect, and then Retire

When onboarding mobile devices with Microsoft Intune, you can utilize other Microsoft 365 security solutions to protect and harden devices. These services are part of Microsoft Defender **Extended Detection and Response (XDR)**. This is a unified security incident platform that integrates multiple security tools and data sources to provide comprehensive threat detection, investigation, and response capabilities across an organization's entire digital environment.

Microsoft Defender for Identity protects against user and sign-in risks for devices that are using Windows Active Directory for identity and access. Cloud-native devices, such as devices running iOS, macOS, and Android, have Microsoft Entra ID Protection for user and sign-in risks. Microsoft Defender for Cloud Apps is a cloud app security broker solution that manages the applications that are being accessed from devices.

Microsoft Defender for Office 365 provides the Defender services that are within the Microsoft 365 suite of products. The Microsoft Defender for Office 365 suite has the following solutions:

- **Microsoft Defender for Office 365**: Protects collaboration services across Exchange Online, SharePoint, OneDrive, and Teams to avoid damaging links and attachments.

- **Microsoft Defender for Cloud Apps**: Protects against shadow IT with cloud and enterprise application monitoring and management, and manages the compliance of cloud applications.

- **Microsoft Defender for Identity**: This is like Microsoft Entra ID Protection. It protects hybrid identity infrastructures with on-premises Windows Active Directory.

- **Microsoft Defender for Endpoint**: Protects and decreases the attack surface for Windows 10/11 devices.

> **Note**
>
> Additional information on these services can be found at this link: `https://learn.microsoft.com/en-us/defender-xdr/microsoft-365-defender?view=o365-worldwide`.

Mobile device protection relies on compliance with company baseline requirements, so MDM and MAM are used to push requirements through compliance profiles. To protect sensitive data and applications from non-compliant devices being accessed, Microsoft Entra Conditional Access policies are created to evaluate these devices for compliance before allowing access. Microsoft Defender for Cloud Apps is also used for access policies through Conditional Access for compliant applications. *Figure 7.6* shows the review flow for Conditional Access policies for users and devices. Conditional Access policies are a helpful solution to protect access to apps and data from mobile devices:

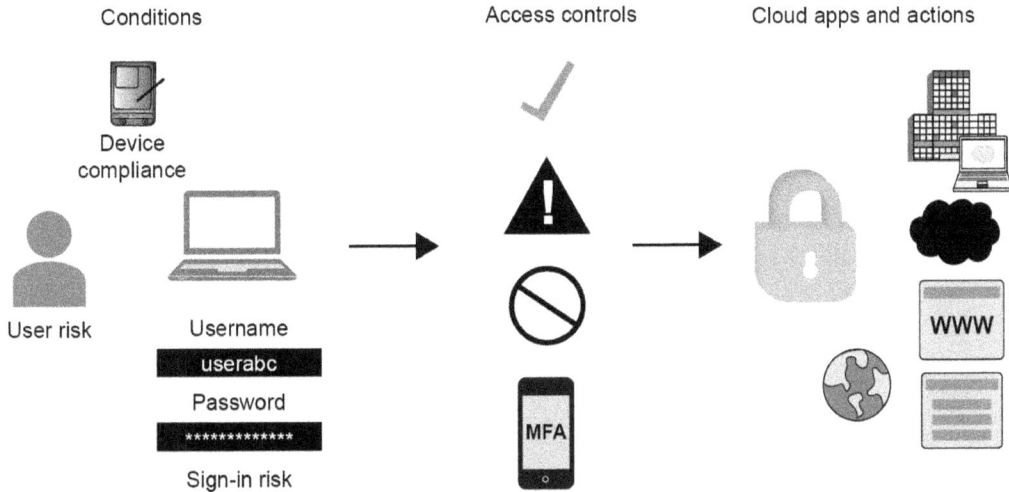

Figure 7.6: Conditional Access policies for MDM-compliant devices, highlighting the various conditions that can be evaluated

Creating and enforcing policies for mobile devices will decrease your attack surface and allow you to harden these devices to avoid exposure to malware, viruses, and unauthorized access to applications.

Educating users about accessing resources with mobile devices is also an important strategy for securing mobile devices. Microsoft Defender for Office 365 has attack simulation tools that provide analysis in training users on common attacks, such as phishing and spear-phishing. You will learn more about this in the *Designing a Strategy for Secure Remote Access* section.

Evaluating Windows LAPS Solutions

Despite the increasingly connected nature of IT systems, there remains a need to have local administrative credentials on systems to ensure that they can be accessed and managed in all scenarios.

These scenarios are often referred to as "break-glass" scenarios and, as the name might suggest, draw on the imagery of breaking the glass on an emergency call point to raise the alarm in the event of an emergency such as a fire.

In the IT world, break-glass scenarios refer to situations such as losing network connectivity to endpoints or access to **identity providers** (**IdPs**). These issues can arise from natural disasters, power outages, cyber-attacks, or misconfigurations. The management of these local administrative credentials can present many challenges:

- Should they be unique to each device? What if you have tens of thousands of such devices?

- Should they be shared across multiple devices? The compromise of a single credential can lead to the unauthorized access of many devices.

- Where and how do you store those local credentials?

- How do you access that location if you need to use those credentials? Is the infrastructure affected by the same event(s) that leads to their need to be used?

- How often do you change those passwords?

How Do You Manage Local Admin Passwords in Windows?

A common practice, for domain-joined Windows devices, was to manage and store these in GPOs – until Microsoft advised (in *Security Bulletin MS14-025* in 2014) that a vulnerability in how those passwords were stored meant that it was possible to retrieve passwords from GPOs that you should not have had access to.

Introduction of Microsoft LAPS

Shortly thereafter, Microsoft introduced **Microsoft LAPS**. This stored the local administrator password for each Windows device in Active Directory, in the computer object for that device.

The password was randomized, unique, and changed regularly by the domain controllers for the domain that the computer was joined to. Access to the passwords was restricted to authorized users using **access control lists** (**ACLs**) and protected in transit using Kerberos v5 authentication and **Asymmetric Encryption Standard** (**AES**) encryption.

There were several limitations of Microsoft LAPS, though:

- It required an extension to the Active Directory schema for the domain(s) you wanted to deploy it on, which requires extensive testing and change control.

- It required the installation of a Group Policy **client-side extension** (**CSE**) agent on each endpoint, which also had dependencies for at least .NET Framework v4 and PowerShell 2.0. This meant it could not be used on operating systems pre-Windows Server 2003 SP1 or Windows Vista SP2.

- It could not manage local administrator credentials for non-domain-joined computers.

- It could not manage other local accounts such as service accounts, scheduled tasks credentials, or SQL local credentials.

- It could not manage devices that were joined to Microsoft Entra.

Replacement of Microsoft LAPS with Windows LAPS

Windows LAPS was introduced by Microsoft in October 2023 and is supported on the following operating system versions or newer:

- Windows 11 22H2 – April 11 2023 update

- Windows 11 21H2 – April 11 2023 update

- Windows 10 – April 11 2023 update

- Windows Server 2022 – April 11 2023 update

- Windows Server 2019 – April 11 2023 update

The benefits of Windows LAPS include the following:

- It can manage the local admin passwords on devices that are joined to AD, joined to Entra, or hybrid-joined devices (joined to both AD and Entra)

- It can back up and manage the **Directory Services Restore Mode** (**DSRM**) account password on Active Directory domain controllers

- It offers protection against common authentication attacks such as pass-the-hash

- Is built into the OS, therefore does not require additional software to be deployed

Deployment Considerations for Windows LAPS

The use of Windows LAPS depends on the type of directory a device is joined to:

Benefits and requirements:

- **Entra joined devices**: Can back up passwords to Microsoft Entra ID

- **Active Directory joined devices**: Can back up passwords to Active Directory

- **Hybrid joined devices**: Can back up passwords to either Microsoft Entra ID or Active Directory

Limitations:

- **Entra joined devices**: Cannot back up passwords to Active Directory

- **Active Directory joined devices**: Cannot back up passwords to Microsoft Entra ID

- **Hybrid joined devices**: Cannot back up passwords to both Microsoft Entra ID and Active Directory simultaneously

Windows LAPS can be configured from Group Policy, from a **configuration service provider (CSP)**, or the CSE agent used by Microsoft LAPS for devices running a supported OS. CSP support allows the configuration to be managed by MDM software, such as Microsoft Intune.

Specifying requirements to Secure AD DS

Many companies have hybrid architectures that still utilize on-premises AD DS as their primary source for managing users, groups, and devices. Protecting identities and maintaining the principles of least privilege for users is as important for AD DS domain controllers as it is for cloud-native identities within Microsoft Entra. Users should not have administrative privileges if they are not necessary to perform their daily tasks. To support administrative control on servers within the domain, you should implement secure administrative hosts to provide hardened systems to control administrative access to other endpoints.

Domain controllers must be secured and hardened against attacks. The domain controllers administer users, groups, and policies across the AD DS architecture. Monitoring, managing, and securing these servers with patch management and security baselines will reduce the attack surface and potential for the domain controllers and member servers on the network to be exposed.

Additional protection within an AD DS infrastructure tied to Microsoft Defender for Office 365 is available with Microsoft Defender for Identity. Microsoft Defender for Identity utilizes the same behavioral analysis as Microsoft Entra ID Protection to find anomalous activities taking place within users and sign-ins. This information is then provided as an alert to Microsoft Defender for Office 365 for security analysts to investigate, along with cloud-native users and applications. *Figure 7.7* shows how sensors on **Active Directory Federated Services (AD FS)** servers and the domain controller provide network traffic and Windows events and traces to Microsoft Defender for Office 365. This two-way communication allows Microsoft Defender for Office 365 to provide additional activity signals that may identify threats to Active Directory:

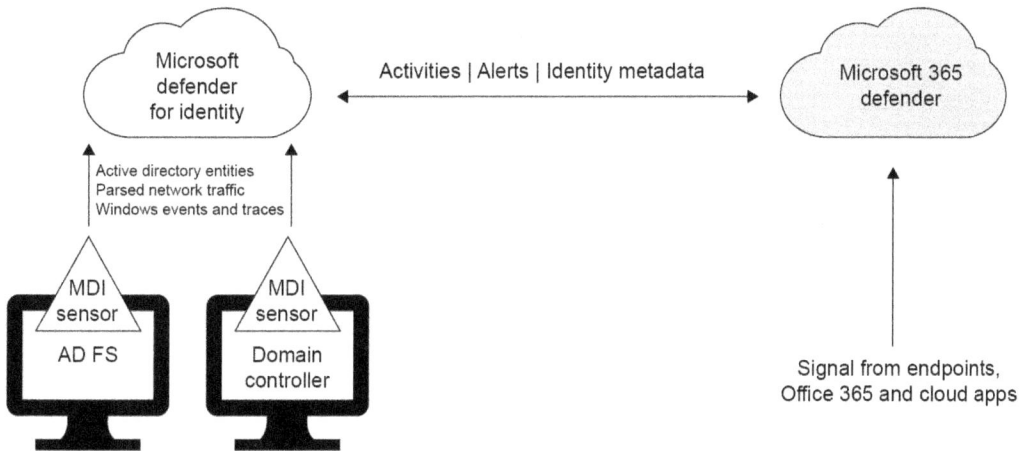

Figure 7.7: Microsoft Defender for Identity (MDI) integrating with Microsoft 365 Defender

Decreasing the attack surface for endpoints includes protecting the identities that have access to those endpoints. Microsoft Defender for Identity provides a solution to protect these identities within a hybrid architecture that utilizes AD DS for user, group, and policy management. The following section will discuss how to design a strategy for managing secrets, keys, and certificates.

Designing a Strategy to Manage Secrets, Keys, and Certificates

Securing secrets, keys, and certificates is as important as securing identities within your cloud and hybrid architectures. Many compliance regulations require customers to manage their keys rather than the cloud provider. Azure allows you to separate these duties with Azure Key Vault. Azure Key Vault is a centralized location where you can protect and manage encryption keys, secrets, and certificates.

Some Azure services use a default configuration for Microsoft-managed keys, such as with Azure storage accounts. *Figure 7.8* shows where you can change that configuration to `Customer-managed keys` within an Azure storage account:

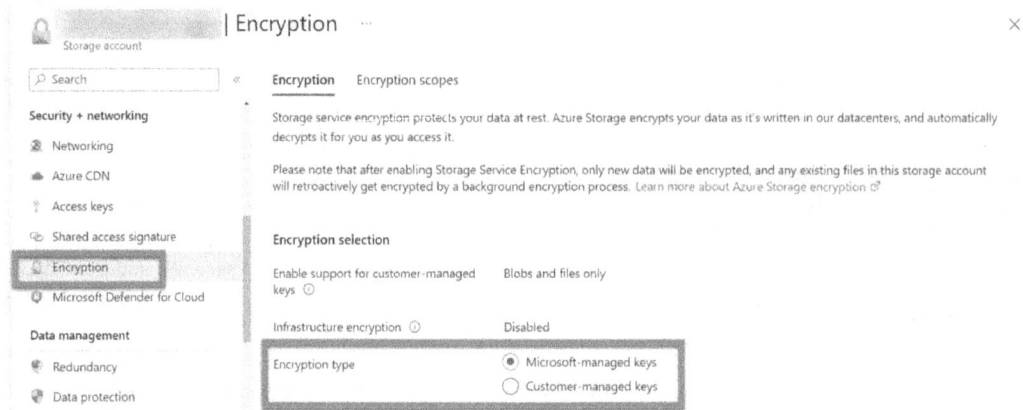

Figure 7.8: Configuring customer-managed keys with Azure Key Vault: a screenshot from the Azure portal

Each of the features of security management is listed here:

- **Secrets management**: Securely store and control access to tokens, passwords, certificates, API keys, and other secrets.

- **Key management**: Azure Key Vault makes it easy to create and control the encryption keys used to encrypt your data.

- **Certificate management**: Here, you can provision, manage, and deploy public and private **Transport Layer Security/Secure Sockets Layer (TLS/SSL)** certificates for use with Azure and your internal connected resources. Azure Key Vault can also be configured to automatically renew these certificates before they expire, maintaining secure availability to websites.

There are two tiers of Azure Key Vault: Standard and Premium. With Premium, you can use Azure Key Vault to synchronize on-premises **hardware security module (HSM)** keys and secrets. Your company may have this as a requirement for full separation of duties and to utilize on-premises HSM keys and secrets within the Azure infrastructure.

Where keys and secrets provide access to resources within Azure and the hybrid infrastructure, certificates allow you to secure and encrypt transmission traffic across intranet and internet sites. Azure Key Vault provides a means to manage the certificates that validate this secure communication. *Figure 7.9* shows how Azure Key Vault provides the communication and validation of certificates with the **certificate authority (CA)** to the cloud application or an on-premises application that is registered to Microsoft Entra:

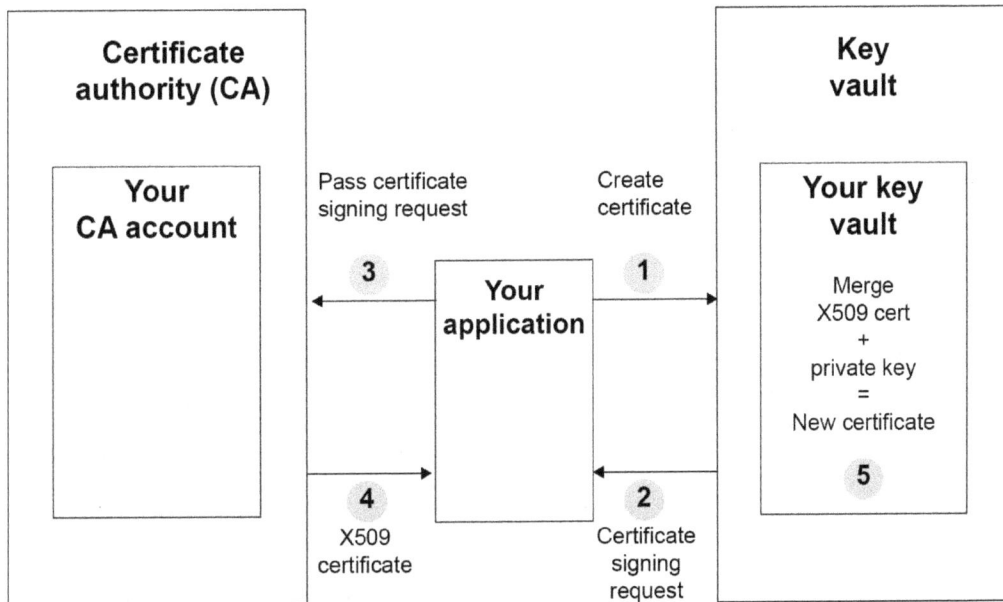

Figure 7.9: CA validation process with Azure Key Vault

In addition to managing and validating certificates for secure communications to your applications, Azure Key Vault can also be used to automate the renewal process of these certificates to prevent potentially exposing the applications to compromised certificates.

In the next section, we will discuss designing a strategy for secure remote access.

Designing a Strategy for Secure Remote Access

When designing for secure remote access, you need to consider this in two separate ways. The first is for remote management of servers and applications, while the second is how mobile devices and users can securely access applications to perform their work tasks.

Remote Management of Servers and Applications

Since you are using resources on Azure and on-premises, you need to consider managing your virtual machines securely without leaving them open to attacks. Typically, you manage either a Linux virtual machine on SSH port 22 or a Windows virtual machine on RDP port 3389. Attackers know this and are known to run programs that check IP addresses for these ports to be open. They can then leverage this for a brute-force attack on your resources. Therefore, it is important not to leave these ports open to the internet.

Azure provides options so that you can avoid having these ports open to the internet, while still making them available to you to manage at the operating system level remotely. The two that will be discussed are **just-in-time** (**JIT**) virtual machine access and Azure Bastion. Azure Arc allows remote management of on-premises servers through SSH for Linux and Windows Admin Center for Windows devices.

JIT virtual machine access provides you with a specific amount of time to access your virtual machine through port 22 or port 3389.

To make a JIT virtual machine access request in Azure, follow these steps:

1. **Open the Azure portal**: Navigate to the Azure portal and go to the virtual machines page.
2. **Select the VM**: Choose the virtual machine you want to access.
3. **Check JIT status**: On the `Connect` page, Azure will check whether JIT is enabled for the VM. If not, you'll be prompted to enable it.
4. **Request access**: If JIT is enabled, click on the `Request access` button.
5. **Specify details**: Select the ports you need access to and specify the time range for access.
6. **Approval and configuration**: Microsoft Defender for Cloud will verify your permissions and configure the necessary **network security group** (**NSG**) and Azure Firewall rules to allow inbound traffic for the specified ports and time.

After the specified time, the access rules will revert to their previous state, ensuring the VM remains secure.

When a JIT virtual machine access request is made, a temporary rule is created in the NSG for that virtual machine network interface that allows the inbound traffic for the port requested from the user's source location. This prevents having a larger attack surface than necessary – the access is only permitted from the IP address of the user requesting the JIT access. *Figure 7.10* shows how this works with the NSG providing an inbound rule for a limited access time:

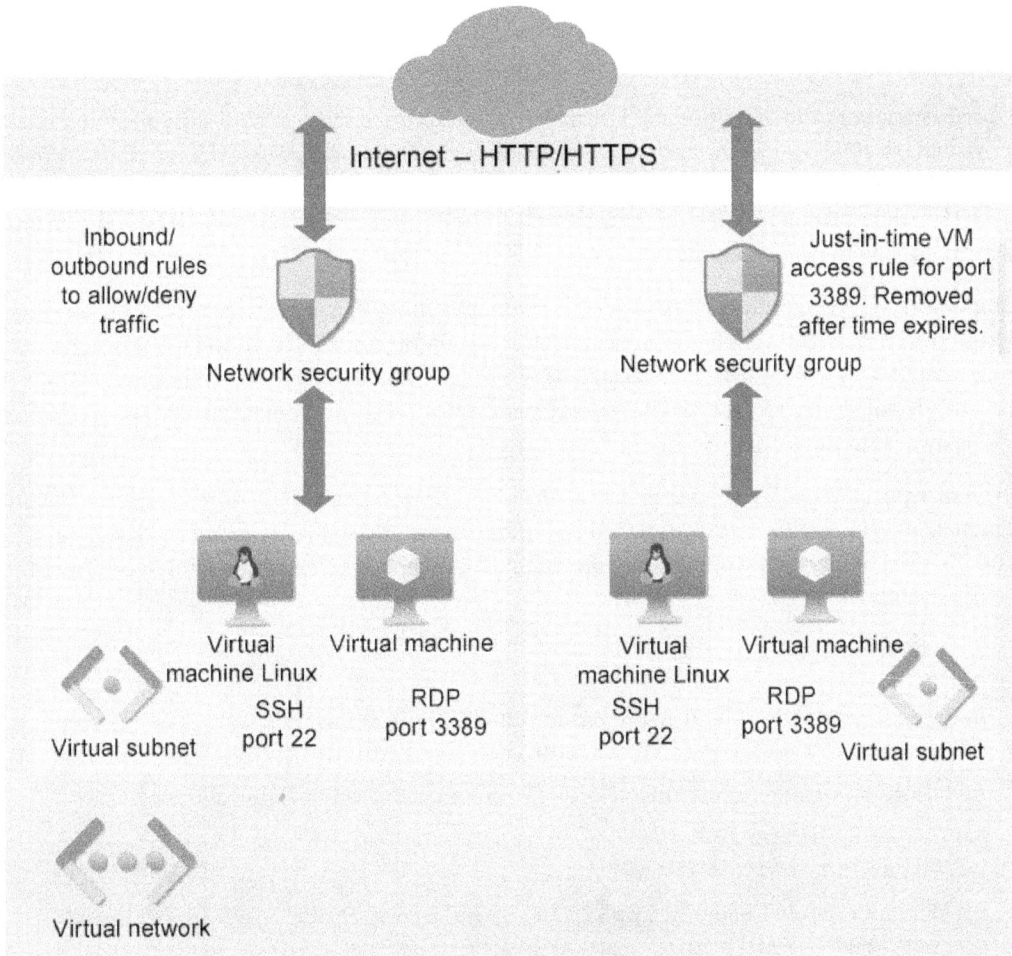

Figure 7.10: NSG with a JIT virtual machine access inbound rule

The maximum amount of time allowed within JIT virtual machine access is set by an administrator within Microsoft Defender for Cloud when JIT virtual machine access is activated.

> **Note**
>
> Additional information on JIT virtual machine access can be found at this link: `https://learn.microsoft.com/en-us/azure/defender-for-cloud/just-in-time-access-usage?tabs=jit-config-asc%2Cjit-request-asc`.

JIT virtual machine access is just one way to provide secure remote management to virtual machines. Next, we will describe secure remote management for virtual machines using Azure Bastion.

Azure Bastion works differently than JIT virtual machine access to manage virtual machines remotely. JIT virtual machine access utilizes the NSG associated with the subnet or network interface to create a temporary inbound allow rule for one of the management ports. Azure Bastion provides additional protection and isolation through the Azure portal and a Bastion subnet to prevent inbound access to these management ports from the internet entirely.

Azure Bastion leverages the ability to access virtual machines through the Azure portal and then opens port 22 or 3389 through a secure connection within the portal. The concept of Azure Bastion is like that of a jump box, where an isolated device is used with a different username and password from the destination device for remote management. An attacker must obtain two sets of usernames and passwords to carry out their attack.

The Azure portal is necessary to connect to virtual machines when using Azure Bastion for secure remote management. Therefore, users who need to perform these management tasks will need to have the **Virtual Machine Contributor** role assigned to them for access to the Azure portal and the virtual machines.

> **Note**
>
> Additional information about Azure Bastion can be found at this link: `https://learn.microsoft.com/en-us/azure/bastion/bastion-overview`.
>
> Additional information about remote access with Azure Arc can be found at these links:
>
> For Linux SSH: `https://learn.microsoft.com/en-us/azure/azure-arc/servers/ssh-arc-overview`.
>
> For Windows Admin Center: `https://learn.microsoft.com/en-us/windows-server/manage/windows-admin-center/azure/manage-arc-hybrid-machines`.

Remote Management of Mobile Devices and Clients

For users and mobile devices that require remote access, you should utilize the strategies discussed in the *Specifying Security Requirements for Mobile Devices and Clients, Including Endpoint Protection, Hardening, and Configuration* section for the MDM and MAM of these devices. Utilizing these strategies and creating compliance profiles allow you to manage how these devices access applications remotely. Conditional Access policies through Microsoft Entra and Microsoft Defender for Cloud Apps can require access to applications and data through managed compliant devices with policies; they can also block locations, and require additional verification for users viewed as at risk.

For additional secure connections, you can create a site-to-site or point-to-site connection from Azure regions to on-premises locations in a hub and spoke configuration. The site-to-site connections will utilize a VPN connection from an Azure VPN gateway to the on-premises appliance, generally a firewall. Point-to-site connections will require users to connect through a VPN application on their mobile devices. MDM on these devices can be used to enforce this connection. *Figure 7.11* shows this hub and spoke configuration to on-premises data centers and remote users:

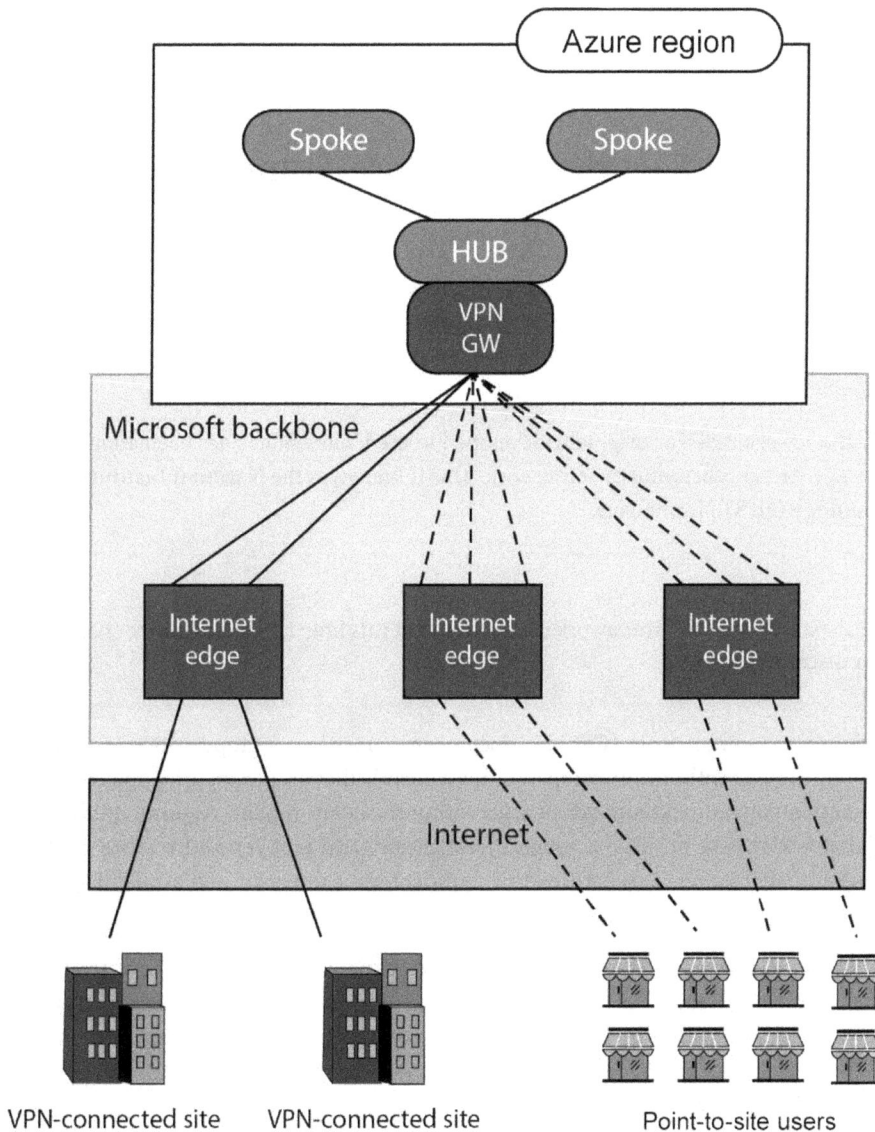

Figure 7.11: Secure connectivity from Azure to on-premises sites

If you want to provide a secure connection that decreases management overhead for users and remote devices, a remote desktop solution can be utilized, such as Azure Virtual Desktop. Azure Virtual Desktop allows you to create a secure and managed desktop image within Azure that users use to access company applications through a secure browser connection to a desktop image within Azure. All company applications and data stay in Azure, including the desktop. When the desktop connection is terminated, no company data or applications remain on the local device, providing a secure access strategy. The attack surface is almost eliminated. The only potential threat is an identity compromise.

The next section will help you understand security operations frameworks and how to perform forensics on endpoints.

Understanding Security Operations Frameworks, Processes, and Procedures

As you build a strategy to secure users, devices, infrastructure, applications, and data, you need to consider the monitoring, management, and operations that will govern the strategy. Documented and maintained security operations frameworks, processes, and procedures are identified and created for this governance.

A popular framework that was originally developed in the United States has been adopted globally by companies as a strong cybersecurity framework. This is known as the **National Institute of Standards and Technology (NIST)** framework.

> **Note**
> Information on the NIST framework can be found at this link: `https://www.nist.gov/cybersecurity`.

NIST's **Cybersecurity Framework (CSF)** has documented practices for proper processes and procedures for security operations within your company. These include that proper security operations should be able to **detect** potential threats and adversaries within the environment, **respond** quickly to identify a potential attack or false positive through investigation, and **recover and restore** workloads and data during and after an attack regarding the confidentiality, integrity, and availability required for those resources.

With any attack, security operations processes should acknowledge and begin investigating the alert quickly. Once an attack has been verified, the time to prevent that attack from affecting additional resources should be reduced by isolating the affected resources from the production network. Preventative measures should be put in place that put additional focus on high-leverage attack areas, such as administrative accounts, and a focus on proactively hunting for threats and reducing the vulnerabilities within the environment.

People are also a key element to successful security operations. Proper education and training through the company can be critical to decrease the potential for attacks. Many attacks are initiated through an internal error, either in a configuration or a compromised user. Educating users on how to recognize a potential attack, such as a phishing email or malicious attachment, will protect against threats to your company's resources.

Some popular metrics should be considered when evaluating the company's security operations' effectiveness. A cybersecurity architect who is brought in to evaluate the effectiveness of the current team should use these metrics to gain insights. These metrics are as follows:

- **Mean time to acknowledge** (**MTTA**) is a metric that identifies the speed at which a security operations team acknowledges an alert at the tier 1 level. This can be the time from the alert being created to a security analyst viewing the incident to identify whether it is a false positive or an actual threat.

- **Mean time to remediate** (**MTTR**) is a metric that identifies how long it has taken to remediate an incident and remove an attacker's access within the environment and the exposure to resources.

- **Incidents remediated** are the number of incidents resolved either manually through the security operations team or with an automated response.

- **Escalations between tiers** track the incidents that have had to be moved to different teams. If incidents are being moved quickly to escalation or escalated to the wrong teams, this could identify a requirement for additional training within the security operations team.

Security operations require an understanding of how to investigate and perform forensic analysis on resources that have been compromised.

Forensic analysis in cybersecurity involves the meticulous examination of digital evidence to identify, investigate, and mitigate security breaches. This process is crucial for understanding the scope and impact of cyber-attacks, preserving evidence for legal proceedings, and improving future security measures.

Microsoft provides tools for security operations to investigate alerts and incidents within Microsoft Defender for Office 365, Microsoft Defender for Cloud, and Microsoft Sentinel. Devices and endpoints that are configured and connected to Microsoft cloud solutions, such as Microsoft Defender for Endpoint, can utilize these solutions to investigate and respond to threats and vulnerabilities.

The next section will provide you with a case study in which you can apply the concepts you have learned in this chapter.

Case Study – Designing a Secure Architecture for Endpoints

Apply what you learned in this chapter by completing the case study on the accompanying online platform. In this case study, you will be given a company scenario and asked to complete several tasks to ensure the company meets the requirements for designing a secure architecture for endpoints.

To access the case study, visit the following link or scan the QR code.

Link to the case study: `https://packt.link/SC100-E2-CaseStudy_Chapter7`

QR code:

Figure 7.12: QR code to access case study for Chapter 7

Summary

In this chapter, we discussed how to design a security strategy for on-premises and cloud virtual machines, as well as other networked endpoints such as mobile and IoT devices. You also learned how to create a secure access strategy and manage keys, secrets, and certificates with Azure Key Vault. This chapter concluded by providing an understanding of the processes, procedures, and people necessary for security operations, and how to evaluate the effectiveness of your security operations.

The key takeaways from this chapter are as follows:

- The need to consider every element of your infrastructure, not just your cloud resources but also the devices that communicate and integrate with it, be they servers, desktop endpoints, mobile devices, or IoT devices

- The need to also consider the security of the communications between all these cloud and non-cloud devices and applications

- The broad set of device types and requirements to support within an IoT/OT/connected devices infrastructure, and how you can leverage Microsoft Defender for IoT to help monitor and secure that infrastructure

- The importance of securely managing privileged local credentials on all devices, and leveraging Windows LAPS to enable you to do that for Windows servers and desktops

- The need to ensure all secrets are securely stored and managed – you can use Azure Key Vault to aid you in this

In the next chapter, you will learn how to design a strategy for securing SaaS, PaaS, and IaaS. This will include building a security baseline for each of these services while covering security requirements for containers, edge computing, application services, databases, and storage accounts within Azure.

Exam Readiness Drill – Chapter Review Section

Apart from mastering key concepts, strong test-taking skills under time pressure are essential for acing your certification exam. That's why developing these abilities early in your learning journey is critical.

Exam readiness drills, using the free online practice resources provided with this book, help you progressively improve your time management and test-taking skills while reinforcing the key concepts you've learned.

How to Get Started

1. Open the link or scan the QR code at the bottom of this page.

2. If you have unlocked the practice resources already, log in to your registered account. If you haven't, follow the instructions in *Chapter 11* and come back to this page.

3. Once you have logged in, click the **START** button to start a quiz.

We recommend attempting a quiz multiple times till you're able to answer most of the questions correctly and well within the time limit.

You can use the following practice template to help you plan your attempts:

Working On Accuracy		
Attempt	Target	Time Limit
Attempt 1	40% or more	Till the timer runs out
Attempt 2	60% or more	Till the timer runs out
Attempt 3	75% or more	Till the timer runs out
Working On Timing		
Attempt 4	75% or more	1 minute before time limit
Attempt 5	75% or more	2 minutes before time limit
Attempt 6	75% or more	3 minutes before time limit

The above drill is just an example. Design your drills based on your own goals and make the most of the online quizzes accompanying this book.

First time accessing the online resources? 🔒

You'll need to unlock them through a one-time process. **Head to** *Chapter 11* **for instructions**.

Open Quiz `https://packt.link/SC100_CH07` Or scan this QR code →	

8

Design a Strategy for Securing SaaS, PaaS, and IaaS

The previous chapter discussed designing a strategy for securing servers and client endpoints. This chapter will discuss designing a strategy for securing **software-as-a-service (SaaS)**, **platform-as-a-service (PaaS)**, and **infrastructure-as-a-service (IaaS)** infrastructures.

Security baselines for operations and governance are key components of managing the security of your infrastructure. Security, networking, and infrastructure administration teams must all work together and have clearly defined roles to manage and maintain security baselines and continued protection for infrastructure, endpoints, identities, and data.

Cloud security requires a different way of thinking than traditional on-premises data center security. The physical security of the building, data center, and network interfaces is no longer "in-house," and IT and security professionals must collaborate on a consistent strategy for implementing, monitoring, and managing infrastructure, devices, identity, and data. This strategy includes the definition of processes and guidelines that will make up the security baselines for the company. In *Chapter 7, Design a Strategy for Securing Server and Client Endpoints*, you learned how to build a strategy for endpoints with an operating system along with on-premises devices.

This chapter targets the domain of *Designing Security Solutions for Infrastructure* in the SC-100 exam guide. It will include building a security baseline for each of these services as well as security requirements for the container, edge computing, application services, databases, and storage accounts within Azure.

In this chapter, you are going to cover the following main topics:

- Specifying security baselines for SaaS, PaaS, and IaaS services

- Specifying security requirements for **Internet of Things (IoT)** workloads

- Specifying security requirements for data workloads, including SQL, Azure SQL Database, Azure Synapse, and Azure Cosmos DB

- Specifying security requirements for web workloads, including Azure App Service

- Specifying security requirements for storage workloads, including Azure Storage

- Specifying security requirements for containers

- Specifying security requirements for container orchestration

- Evaluating solutions that include Azure **Artificial Intelligence (AI)** services security

- Case study – security requirements for IaaS, PaaS, and SaaS

Specifying Security Baselines for SaaS, PaaS, and IaaS Services

This section will discuss building a security baseline and security requirements for cloud services that are SaaS, PaaS, and IaaS.

Microsoft **Cloud Adoption Framework (CAF)** for Azure provides a roadmap for the planning and adoption of cloud services.

> **Note**
>
> Security guidelines for the CAF can be found here: `https://learn.microsoft.com/en-us/azure/cloud-adoption-framework/secure/`.
>
> It is vitally important to read through and understand the CAF material – it will come up in the exam.

Figure 8.1 shows the CAF process for cloud adoption with continuous security, management, and governance, providing a cyclical process:

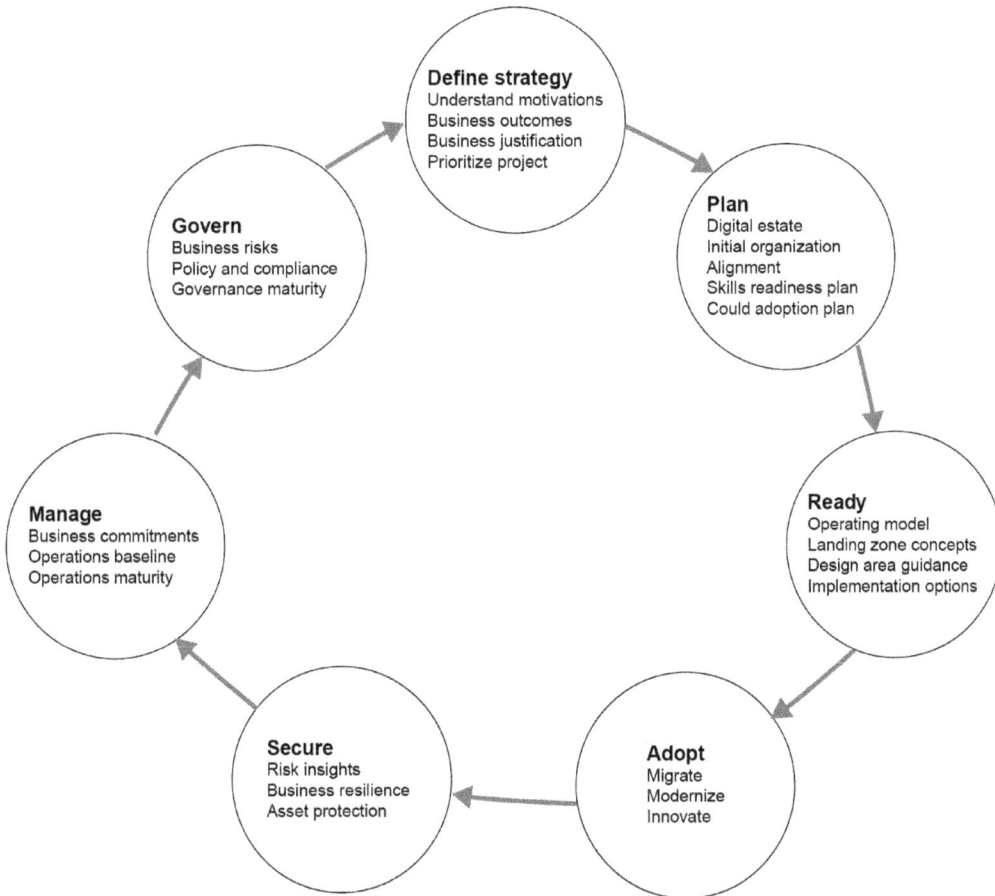

Figure 8.1: CAF

The bottom part of the CAF outlines the shift-left approach to continuously improve the security posture and review the environment for risks and vulnerabilities. The shift-left approach to security is continuous governance of compliance with security and policies. This includes utilizing tools and solutions for **cloud security posture management (CSPM)**, **cloud workload protection platforms (CWPPs)**, and **cloud-native app protection platforms (CNAPPs)**. Each cloud service has a different level of security responsibility shared between the customer company and the cloud provider, Microsoft. Meticulously planning the team's roles and responsibilities and utilizing the CAF for quick wins along the cloud security journey will build confidence and create a solid foundation going forward.

As a review, *Table 8.1* provides the shared responsibility model for on-premises, IaaS, PaaS, and SaaS:

Responsibility	On-premises	IaaS	PaaS	SaaS
Data governance and rights management	Customer	Customer	Customer	Customer
Client endpoints	Customer	Customer	Customer	Customer
Account and access management	Customer	Customer	Customer	Customer
Identity and directory infrastructure	Customer	Customer	Microsoft/Customer	Microsoft/Customer
Application	Customer	Customer	Microsoft/Customer	Microsoft
Network controls	Customer	Customer	Microsoft/Customer	Microsoft
Operating system	Customer	Customer	Microsoft	Microsoft
Physical hosts	Customer	Microsoft	Microsoft	Microsoft
Physical network	Customer	Microsoft	Microsoft	Microsoft
Physical data center	Customer	Microsoft	Microsoft	Microsoft

Table 8.1: Shared responsibility for Microsoft security

Note in *Table 8.1* that full responsibility from top to bottom is the burden of the customer for on-premises. This responsibility shifts more to sharing more responsibility with the cloud provider as you move to IaaS, PaaS, and SaaS infrastructure.

However, as a cybersecurity architect, you should be aware of the security capabilities in place to protect identities and data. Even though Microsoft provides many of the security controls as a baseline, you, as the customer, should understand the level that these controls protect. Additional enhanced controls are required for networking and applications in PaaS, and identity and directory infrastructure for PaaS and SaaS.

The next few sections will help you address these baselines and responsibilities as a cybersecurity architect. The first service that we will discuss is SaaS.

Security Baselines for SaaS

We will start with SaaS because it has the least amount of responsibility for the customer to maintain a secure environment. SaaS applications have been developed to provide an end-user experience without any need to configure the infrastructure. Only you can determine how data, identities, and endpoint devices are protected. The identity and directory infrastructure are provided as a baseline by Microsoft. You are responsible for determining whether or not additional controls are required to protect that infrastructure, such as **multi-factor authentication** (**MFA**), identity protection, password protection, and conditional access. These controls also feed into the account and access management of a SaaS application.

Protecting of client endpoints is an especially important part of a SaaS infrastructure. While SaaS applications are safeguarded against physical threats through hardening, compromised client endpoints infected with malware, viruses, or other malicious code can disrupt performance. Therefore, it is essential to develop and implement a strategy that addresses the requirements and baselines for these client endpoints. *Chapter 7, Design a Strategy for Securing Server and Client Endpoints*, specified how the Security Compliance Toolkit can be used on Windows endpoints to create a baseline of protection. Microsoft Defender for Endpoint will also help to harden the endpoints and decrease the attack surface. For the protection of Microsoft 365 SaaS applications, there is Microsoft Defender for Office 365. Microsoft Defender for Office 365 has tools to protect against malicious links, attachments, and phishing. This includes a helpful attack simulator that can be used for understanding and training users in identifying threats. Microsoft Defender for Office 365 is one of the solutions within the Microsoft Defender suite to provide a CWPP for SaaS applications.

Microsoft Defender for Cloud Apps helps you to evaluate SaaS applications and registered applications that are being accessed on your network that may be viewed as suspect and could compromise your security and compliance. For all cloud services, there are tools within Microsoft Entra, such as MFA, Conditional Access policies, Microsoft Entra Identity Protection, and Microsoft Entra Password Protection. These tools are utilized to identify threats, remediate vulnerabilities within the identity, and help users access applications.

Figure 8.2 shows the discovery dashboard for Microsoft Defender for Cloud Apps, which allows you to manage and monitor applications with CSPM:

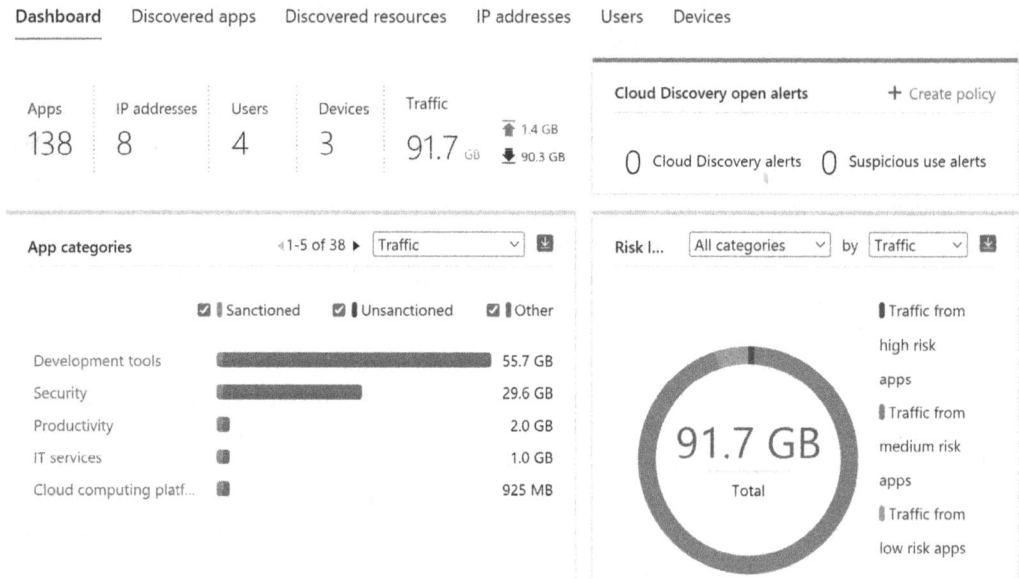

| Dashboard | Discovered apps | Discovered resources | IP addresses | Users | Devices |

Apps	IP addresses	Users	Devices	Traffic
138	8	4	3	91.7 GB ⬆ 1.4 GB ⬇ 90.3 GB

Cloud Discovery open alerts   + Create policy

O Cloud Discovery alerts   O Suspicious use alerts

App categories   ◄1-5 of 38 ► Traffic ▾ ⬇

☑ | Sanctioned   ☑ | Unsanctioned   ☑ | Other

Development tools	55.7 GB
Security	29.6 GB
Productivity	2.0 GB
IT services	1.0 GB
Cloud computing platf...	925 MB

Risk l... All categories ▾ by Traffic ▾ ⬇

91.7 GB
Total

| Traffic from high risk apps
| Traffic from medium risk apps
| Traffic from low risk apps

Figure 8.2: The Microsoft Defender for Cloud Apps discovery dashboard

In the dashboard, accessed apps and traffic are analyzed and categorized into high-, medium-, and low-risk levels. This allows you to review what users are accessing and determine whether any applications may be a threat to your company when being accessed.

> **Note**
>
> For more information on Microsoft Defender for Cloud Apps, see this link: `https://learn.microsoft.com/en-us/defender-cloud-apps/what-is-defender-for-cloud-apps`.

There are limits to the number of security controls that you can put in place within a SaaS application, which requires more due diligence and care to determine the proper baseline of controls you have control over.

The same areas of responsibility and control that apply to SaaS can also apply to PaaS and IaaS. In the next section, you will learn about the additional areas of controls and responsibilities for creating an IaaS security baseline.

Security Baselines for IaaS

As discussed in the previous section, SaaS has a limited level of control you can enable for protection. However, understanding and utilizing those controls properly to create a baseline for security are all part of due care and due diligence. IaaS is the other end of the spectrum, where you are responsible for protecting all areas above the physical network infrastructure, physical host, and building. *Chapter 7, Design a Strategy for Securing Server and Client Endpoints*, provided the security baseline guidance for the virtual machine, or server, operating systems. These operating systems will be either Windows or Linux in an Azure virtual machine infrastructure.

These virtual machines and applications within an IaaS infrastructure need to be protected from malware and viruses. The data and virtual machine disks should be encrypted at rest with Azure Disk Encryption and storage accounts with Storage Service Encryption. Transmission of data should utilize encrypted transmission with **Secure Socket Layer (SSL) / Transport Layer Security (TLS)**, or a dedicated secure connection, such as VPN or ExpressRoute.

Microsoft Defender for Cloud can be used to continuously assess resources on Azure and provide recommendations for improvements of the security baseline in the Azure Security Benchmark. Microsoft Defender plans provide CWPPs and alerts regarding potential vulnerabilities and threats on the workloads.

Figure 8.3 shows the Microsoft Defender for Cloud security posture improvement recommendations for IaaS resources:

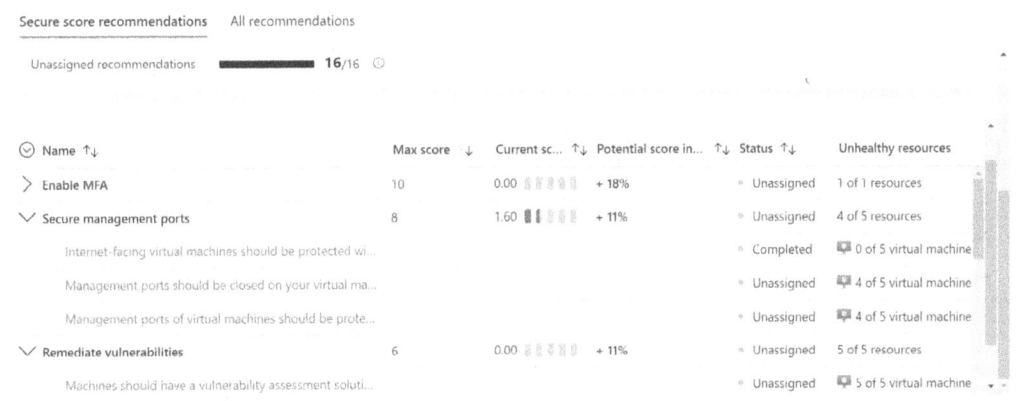

Name ↑↓	Max score ↓	Current sc... ↑↓	Potential score in... ↑↓	Status ↑↓	Unhealthy resources
> Enable MFA	10	0.00 + 18%		Unassigned	1 of 1 resources
∨ Secure management ports	8	1.60 + 11%		Unassigned	4 of 5 resources
Internet-facing virtual machines should be protected wi...				Completed	0 of 5 virtual machine
Management ports should be closed on your virtual ma...				Unassigned	4 of 5 virtual machine
Management ports of virtual machines should be prote...				Unassigned	4 of 5 virtual machine
∨ Remediate vulnerabilities	6	0.00 + 11%		Unassigned	5 of 5 resources
Machines should have a vulnerability assessment soluti...				Unassigned	5 of 5 virtual machine

Figure 8.3: Microsoft Defender for Cloud virtual machine security recommendations

> **Note**
>
> Additional information on Microsoft Defender for Cloud and protection for IaaS resources with Microsoft Defender for Servers can be found at this link: `https://learn.microsoft.com/en-us/azure/defender-for-cloud/plan-defender-for-servers`.

Security Baselines for PaaS

PaaS resources fall in between IaaS and SaaS for responsibility. Microsoft provides many security controls and baselines for protecting the infrastructure, including hardening the operating system. However, for networking and application protections, additional controls are available that you should evaluate. These controls include adding Azure Front Door or Application Gateway with a **web application firewall (WAF)**. *Figure 8.4* compares and shows where some of these controls come into use:

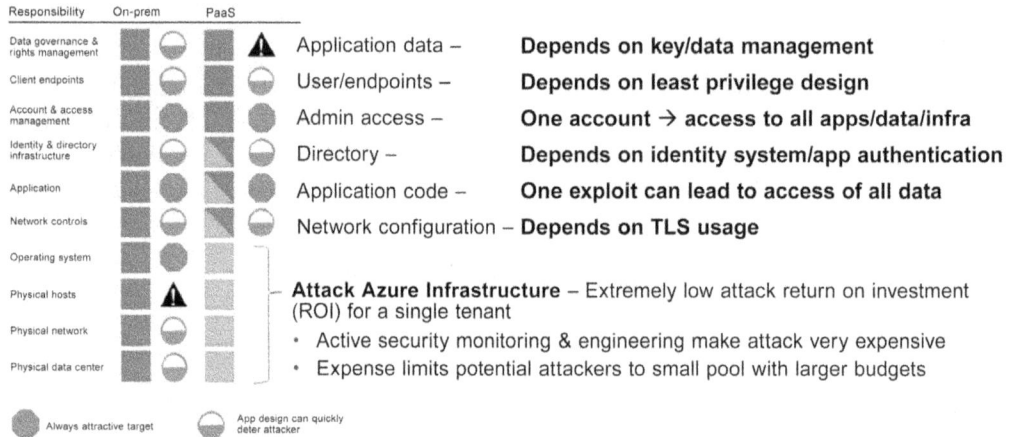

Figure 8.4: Security controls for PaaS

Figure 8.5 shows how shared responsibility comes into play for PaaS applications and where vulnerabilities and threats could be exposed without certain controls in place. The Azure Security Benchmark is a helpful tool for evaluating applications on PaaS resources within Microsoft Defender for Cloud.

> **Note**
>
> For more information, use this link: `https://learn.microsoft.com/en-us/security/benchmark/azure/baselines/app-service-security-baseline`.

Microsoft Defender for Cloud's security posture and Secure Score recommendations provide ways to improve your company's alignment with the Azure Security Benchmark and create a strong security baseline for your PaaS resources.

As you continue this chapter, you will learn about additional requirements for securing various PaaS resources.

Specifying Security Requirements for IoT Devices and Connected Systems

IoT and connected systems have continued to increase in prevalence within our technological society. An **IoT device** is any device connected to a private or public network. These are generally operational technologies that collect data that can be analyzed, such as a thermostat, machine sensor, or traffic camera. In many cases, these devices send data over insecure networks. Securing these devices requires an amount of due diligence and understanding of what data is being collected and how.

Evaluating and determining the security requirements for IoT workloads should be divided into the different components that make up an IoT infrastructure. These areas are as follows:

- Device
- Field gateway
- Cloud gateway
- Services

Each of these areas becomes a potential attack surface and should be reviewed and treated as a security boundary to be isolated with zero-trust principles. Microsoft Defender for IoT can be used to provide security posture management and threat detection on the IoT infrastructure. *Figure 8.5* shows the workflow for Defender for IoT protecting an IoT infrastructure:

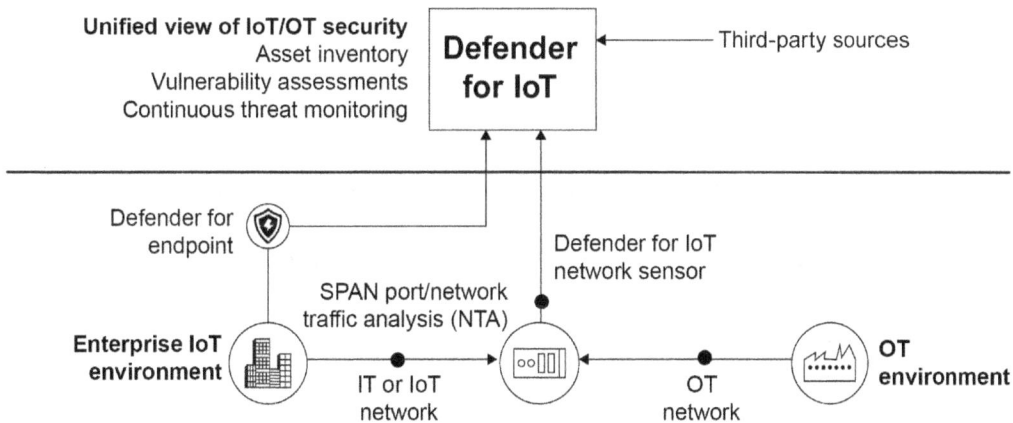

Figure 8.5: Defender for IoT threat detection diagram

Defender for IoT protects IoT infrastructures for **operational technology** (**OT**) and enterprise IoT environments. Enterprise IoT devices are generally on a more open network architecture than OT devices. Still, both are susceptible to threats if they are not protected with proper zero-trust access principles. Microsoft Defender for IoT is currently a separate service from other Microsoft Defender for Cloud plans. *Figure 8.6* shows Defender for IoT getting started with the two plans available for OT and enterprise IoT:

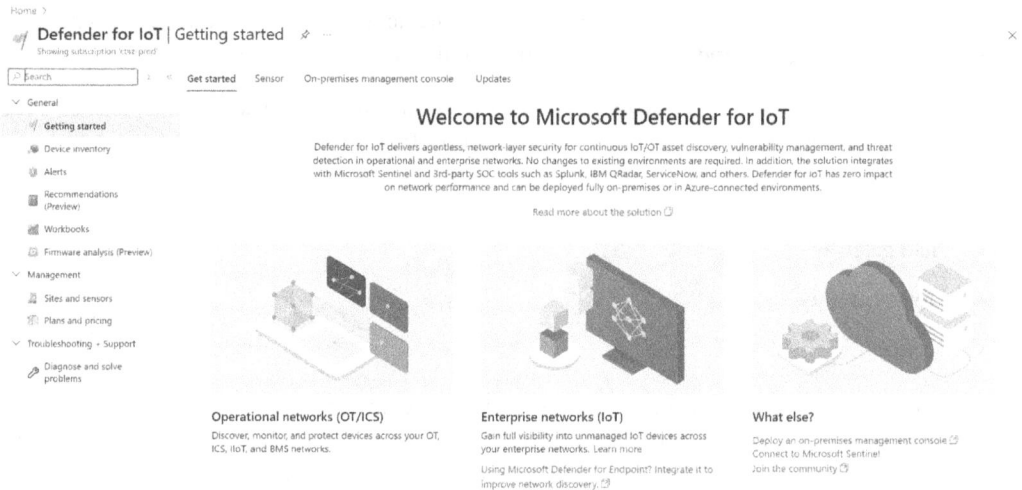

Figure 8.6: Microsoft Defender for IoT getting started

IoT and connected systems are a challenge for many organizations to secure due in large part to the volume of devices, the variety of software and hardware configurations, and the inherent limitations of each of these factors.

While a start-up company may have the opportunity to standardize on a core set of standard IoT devices and software, enterprises undergoing a cloud transformation journey where there is already a sizeable estate of existing devices may find this particularly challenging to integrate into their cloud estate.

There are several considerations to factor into your IoT and cloud configuration and deployment and these are discussed in more detail in the following sections.

Device Security

To ensure the most secure device possible, you should seek to do the following:

- Reduce the attack surface by ensuring there are no hardware components in the device that are not essential to its operations, such as USB ports, network ports, and so on.

- Ensure that the device uses security features, such as encryption and **trusted platform module (TPM)** chips, to ensure the data is protected, as is the boot process itself.

  A TPM is a specialized microchip designed to enhance the security of your computer. Here are some key points about a TPM:

  - **Security functions**: A TPM primarily handles security-related tasks, such as generating, storing, and managing cryptographic keys. This helps protect sensitive data and ensures that the system's integrity is maintained.

  - **Encryption**: It works with encryption software such as BitLocker to encrypt your hard drives, ensuring that your data remains secure even if your device is lost or stolen.

  - **Authentication**: A TPM can be used with authentication systems such as Windows Hello to securely store biometric data and ensure that only authorized users can access the device.

  - **System integrity**: A TPM can verify the integrity of the system by checking the firmware and operating system for any signs of tampering. This helps prevent unauthorized changes to the system.

  - **Hardware integration**: Typically, a TPM is a separate chip on the motherboard, but it can also be integrated into the main processor in modern systems.

- Ensure that the device cannot be tampered with and that any such attempts invoke appropriate responses such as alerting, device lockdown, and so on.

- Ensure that authentication (to the device and from the device) uses secure methods.

- Ensure that the device can be securely and frequently updated with the latest software and firmware fixes.

- Use **software development kits (SDKs)** where possible – it is usually best not to write your own encryption or authentication code, for example. Use tried and trusted SDKs for that and focus on the business logic in your code.

- Ensure that secrets (passwords, certificates, keys) are always securely stored and accessed and remain confidential.

Following these best practices will provide the best chance of ensuring the security of your data and the availability of your systems and services.

Connection Security

As secure as you might make the device itself, most IoT devices do not exist to operate in isolation and will require some form of connectivity to be able to report data to other systems or receive updates from them (in the case of Azure, using IoT Hub or IoT Central).

This connectivity, though essential, is another attack path and so must be secured using the following:

- X.509 **certificate-based authentication (CBA)** is recommended as a stronger option than usernames and passwords, which are generally cryptographically weak.

- TLS v1.2 or higher (older versions are now less secure and as such are deprecated and not recommended for use).

- Ensure that you have considered the certificate life cycle for all TLS certificates on the device, especially the root certificate. Ensure that you have notifications in place for soon-to-expire certificates and a well-tested method to deploy new certificates and ensure successful rollover from old to new certificates.

- Where possible, enable the Private Link functionality to provide private access to your Azure virtual networks and avoid them being accessible over the public internet.

Cloud Security

Now that we have considered the security of the device and its connectivity, we must focus on the cloud resources that can manage the device and/or that receive data from it.

Many of these considerations are fundamental security best practices, which should not be a surprise to a security professional, and include the following:

- Ensuring the protection of credentials in the cloud by monitoring their use, regularly rotating the credentials, securely vaulting them, and ensuring they are not publicly accessible. This extends to credentials at all stages of the **software development life cycle (SDLC)**, including those used to access or configure deployments, access applications, and transfer data.

- Ensuring the adoption of the principle of least privilege when defining role-based access control in all elements of the solution, using the most secure authentication mechanism available for the solution you are deploying – either a **shared access signature (SAS)** or a Microsoft Entra identity, such as a service principal for IoT Hub or Microsoft Entra identities or authentication tokens for IoT Central.

- In the same way that you should reduce the attack surface of hardware by ensuring there are no unnecessary hardware components in the device, you should also ensure that you understand and minimize software dependencies that may expose you to vulnerabilities. Understand and minimize API endpoint access and the software libraries in use to the bare minimum required for the solution to function.

- Set up diagnostics, monitoring, and alerting to ensure that you have insights into the behavior of your devices and applications and any unusual behavior, as near to real time as possible for your solution architecture.

To recap, Azure IoT Central affords control over the following components from a security configuration perspective:

- Access to your application by users, devices, and programmatic interfaces

- Authentication from your application to other services

- Use of a secure virtual network

- Audit logging to monitor your application

Evaluating Solutions for Securing OT and Industrial Control Systems (ICSs) by Using Microsoft Defender for IoT

Microsoft Defender for IoT (**MDIoT**) can help you with many of these recommendations. The Azure portal displays recommendations for your IoT applications and devices, such as hardware/firmware with known **Common Vulnerabilities and Exposures** (**CVEs**), unauthorized devices, weak passwords, and insecure communications protocols.

It can do this via insights that it gathers from your IoT devices through either a micro agent installed on the device or through agentless monitoring.

The diagram in *Figure 8.7* illustrates the components and communication flows involved for the micro agent.

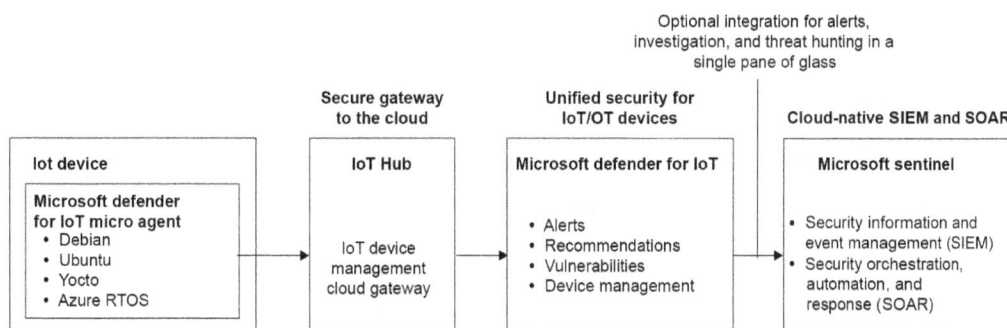

Figure 8.7: Microsoft Defender for IoT micro agent architecture

You can see that the agent can be installed on many Linux operating systems (such as Debian and Ubuntu) as well as IoT-specific operating systems.

The agent sends events and metadata to IoT Hub in your Azure subscription, which acts as a cloud gateway device. This then forwards that information to MDIoT for it to generate alerts, recommendations, lists of vulnerabilities, and device health information.

This in turn can be integrated with Microsoft Sentinel for SIEM and SOAR capabilities, just as with any other component in your Azure infrastructure.

While some devices support the installation of MDIoT security agents that perform monitoring and integrate with Microsoft Defender and Microsoft Sentinel, for those that do not, Microsoft also offers an agentless capability, which is useful when you have legacy devices that cannot support a security agent.

It also enables you to discover IoT devices that you may be unaware of and are not recorded in your **configuration management database (CMDB)**.

Agentless scanning is enabled through the deployment of network sensors that capture and analyze the traffic between your devices and your cloud infrastructure, and are available in two variants.

Cloud-Connected Sensors

Cloud-connected sensors are connected to Microsoft Defender for IoT in Azure, affording the ability to automatically upload the sensor data to Azure in near real time (dependent upon connectivity) and to also automatically receive threat intelligence updates from Microsoft, which dynamically update the sensor capabilities to detect and alert on new/emerging threats and **indicators of compromise (IoCs)**.

Local OT Sensors

Local OT sensors, on the other hand, are managed locally, either directly from the sensor console or infrastructure in the same on-premises location or network, and require a local management server (a management server is required to view the data from multiple sensors in one location rather than directly on the console for each sensor).

Similarly, threat intelligence updates from Microsoft must be manually uploaded to the sensors.

This is a common deployment option for secure facilities with air-gapped networks such as nuclear power plants and other critical national infrastructure, where allowing public access to those networks from the internet would be inadvisable due to the potential consequences of those networks and devices being compromised by malicious actors.

> **Note**
>
> For further in-depth information about agentless monitoring, please review the documentation from Microsoft at this link: `https://learn.microsoft.com/en-gb/azure/defender-for-iot/organizations/overview`.
>
> For further in-depth information about MDIoT security agents, please review the documentation from Microsoft at this link: `https://learn.microsoft.com/en-gb/azure/defender-for-iot/device-builders/overview`.

Specifying Security Requirements for Data Workloads, Including SQL, Azure SQL Database, Azure Synapse, and Azure Cosmos DB

Data is the second-most important asset within your company after people. Data not only includes business and financial data but also a large amount of **personally identifiable information** (PII) regarding employees and customers. If you are a healthcare company, you will also have **personal health information** (PHI). Exposure of this data, whether malicious or unintentional, will cause damage to your company, customers, and people.

Determining security requirements and protecting data workloads should focus on three key criteria:

- Data protection should be in place across all data workloads through the identification of sensitive data and classifying that data with labels to mask data from unauthorized users

- Prevent data loss by utilizing **data loss prevention** (DLP) tools that identify and protect sensitive data that may be accidentally or intentionally shared

- Least privilege access should be enforced by providing the permissions necessary for users to perform their daily work duties

By steadfastly adhering to these three criteria, you can significantly reduce the risk of data exposure to attacks.

In this context, Microsoft Purview plays a crucial role by offering compliance and governance across various data workloads, including Microsoft 365, Azure, SaaS applications, and on-premises data solutions. *Figure 8.8* illustrates the range of data services supported by Microsoft Purview:

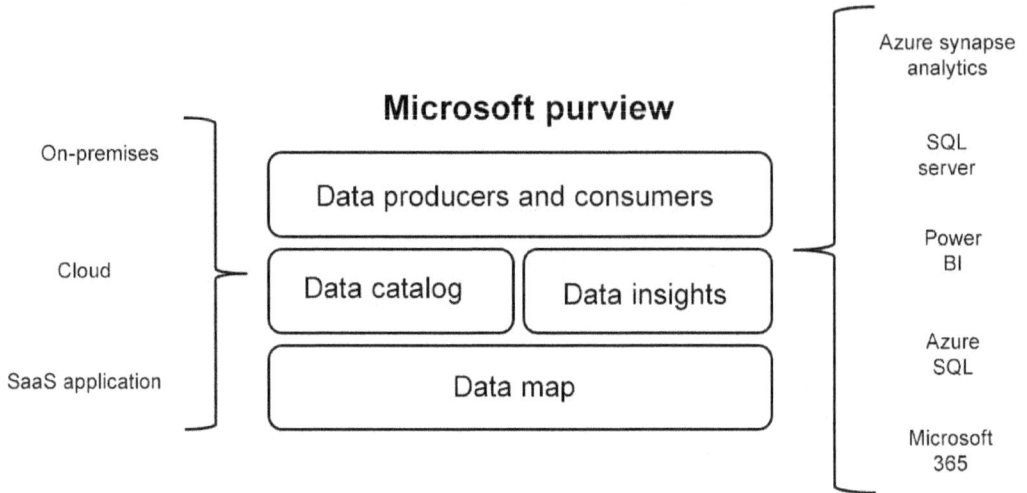

Figure 8.8: Microsoft Purview data sources

Data within companies resides in many separate locations. These locations include storage servers on-premises, cloud storage services (such as OneDrive), storage sites (SharePoint and Teams), SaaS applications (Microsoft 365), and databases.

You can utilize Microsoft Fabric as a solution to govern and use data.

Microsoft Fabric is an end-to-end analytics and data platform designed to unify various data-related tasks within a single, cohesive environment. It integrates data movement, processing, ingestion, transformation, real-time event routing, and report building into one platform.

Built on a SaaS foundation, Microsoft Fabric combines components from Power BI, Azure Synapse Analytics, Azure Data Factory, and more. This integration allows for seamless data management, governance, and analytics, making it easier for enterprises to derive actionable insights from their data.

Microsoft Fabric integrates seamlessly with Microsoft Purview to provide comprehensive data governance and compliance capabilities. Here's how they work together:

- **Unified data governance**: Microsoft Purview allows you to discover, classify, and manage Microsoft Fabric items within its applications. This integration ensures that all data within Fabric is governed under the same policies and standards.

- **Data Catalog**: The Microsoft Purview Data Catalog automatically captures metadata about your Microsoft Fabric items, providing a live view of your data assets. This helps in maintaining an up-to-date map of your data estate.

- **Information Protection**: Sensitivity labels from Microsoft Purview Information Protection can be applied to Fabric data, ensuring that sensitive information is classified and protected. These labels remain intact even when data is exported through supported paths.

- **Compliance and audit**: Microsoft Purview's compliance and audit capabilities extend to Microsoft Fabric, allowing administrators to monitor activities and ensure regulatory compliance across the entire data estate.

- **Centralized management**: The Microsoft Purview hub within Microsoft Fabric provides a centralized page for administrators and users to manage and govern their Fabric data estate. It offers insights into data sensitivity, item endorsement, and links to advanced governance capabilities.

Platform database services are the backend storage sources for most applications developed in the cloud. Azure has many PaaS databases, including Azure SQL Database, Azure Cosmos DB, and Azure Synapse Analytics. Security controls for these database services can be evaluated through Microsoft Defender for Databases. The Microsoft Defender services continuously evaluate and assess security controls against the Azure Security Benchmark and provide recommendations to increase the security posture and identify additional security requirements. Microsoft Defender for Databases allows you to select the PaaS database resources you utilize within your Azure subscription.

> **Note**
>
> For more information on the many different supported data sources, both within and outside of the Microsoft ecosystem, go to this link: `https://learn.microsoft.com/en-us/purview/microsoft-purview-connector-overview`.
>
> For more information on enabling the different Microsoft Defender for Databases plans, go to this link: `https://learn.microsoft.com/en-us/azure/defender-for-cloud/tutorial-enable-databases-plan`.

Figure 8.9 shows some of the security improvement recommendations you may find in Microsoft Defender for Cloud when utilizing Azure SQL Database:

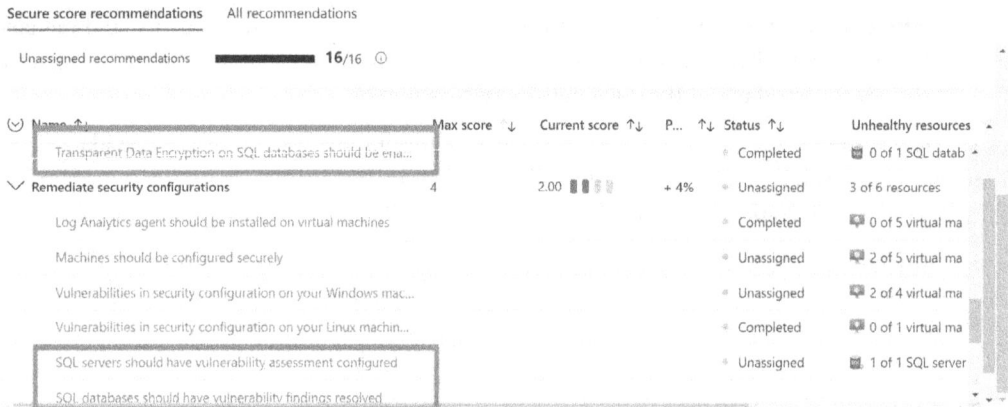

Figure 8.9: Microsoft Defender for Cloud SQL security recommendations

Specifying Security Requirements for Storage Workloads, Including Azure Storage

Protecting data and the data life cycle is necessary for all companies utilizing cloud technologies. As discussed in the *Specifying Security Requirements for Data Workloads, Including SQL, Azure SQL Database, Azure Synapse, and Azure Cosmos DB* section of this chapter, when discussing securing databases, specifying the proper security requirements is a responsibility that a cybersecurity architect should define for engineers and administrators. Among these requirements, you must include, but are not limited to, the following settings and security features:

- Within Azure Storage, you should turn on soft delete for blob data and file shares. This protects data from accidental deletion and allows deleted items to be recovered.

- Azure Storage provides options to use storage account access with Azure **Role-Based Access Control (RBAC)**, but the preferred method would be to utilize Microsoft Entra permissions to access storage account data. Using Microsoft Entra to access storage accounts will help you analyze activity for anomalous behavior that can alert you to potential malicious and brute-force behavior.

- When providing access to storage accounts and data, the principle of least privilege should be applied to assigning permissions. This includes service principals for resources to access storage account data.

- For business-critical data, blob versioning or immutable blobs should be used.

- Anonymous or default internet access to storage accounts should not be used. Misconfigured access creates a potential network vulnerability for the data.

- Separation of public and private data should be exercised within storage accounts. Sensitive data should not be stored in the same storage accounts as publicly accessed data.

- Firewall rules should be used to limit access to your storage account, and network access to storage accounts should be limited to specific networks.

- Only trusted Microsoft services should be allowed access to storage accounts.

- Data should only be transmitted over encrypted channels by enabling secure transfer on storage accounts with HTTPS connections.

- When using SASs, they should also be limited to connections through HTTPS.

- SAS access should have time-bound access and expire within a reasonable period to provide access to the user or resource while limiting the time that that information could be exposed if that SAS path is compromised.

- SAS should be used to provide non-Microsoft Entra users and resources access to storage accounts. Shared key authorization should not be used. When possible, customer-managed, not Microsoft-managed, keys should be used to protect and encrypt storage account data at rest.

- Customer-managed and Microsoft-managed keys should be rotated regularly. When using Azure Key Vault, this can be done with key rotation policies.

Microsoft Defender for Cloud's security posture management provides recommendations for protecting storage accounts based on the Azure Security Benchmark. Microsoft Defender for Storage can be turned on for additional protection of storage accounts. *Figure 8.10* shows a diagram of the workflow of Microsoft Defender for Storage to protect storage accounts:

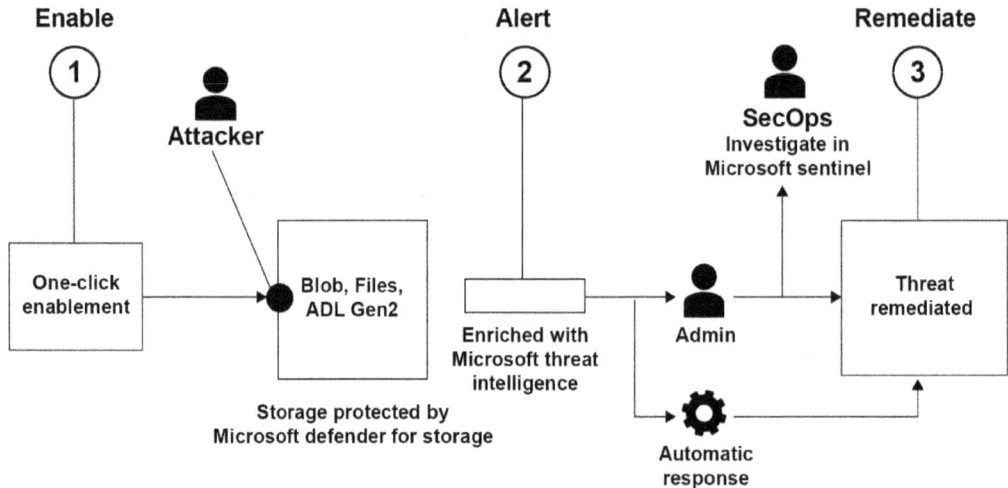

Figure 8.10: Microsoft Defender for Storage threat protection

Microsoft Defender for Storage utilizes Microsoft Threat Intelligence to identify potential security threats that trigger an alert to administrators and an automated response to block and remediate the threat.

> **Note**
>
> More information on Microsoft Defender for Storage can be found at this link: `https://learn.microsoft.com/en-us/azure/defender-for-cloud/defender-for-storage-introduction?tabs=azure-security-center`.

Specifying Security Requirements for Web Workloads, Including Azure App Service

Many storage accounts and data workloads have an application in front of them that accesses and utilizes that data. This may be for an internal application for employees to access company information or it could be a web application for the company website where the public views and purchases company products. In either situation, the web workloads should be secured and access to data should be protected. Some security requirements that should be recommended are as follows:

- Apps should use secure encrypted transmission through HTTPS, which is a default setting within Azure App Service. You should disable any HTTP transmission and accessibility. Azure Key Vault can be used to manage the TLS certificates and renew them before they expire.

- Azure App Service should create static IP restrictions to only allow requests from trusted subsets of IP addresses. By default, App Service accepts requests from any IP address on the internet, so it is recommended to implement additional security measures for secure connections.

- Access to Azure App Service should utilize and enable authentication and authorization for client access to web apps. Application secrets should be private and stored in Azure Key Vault. Secrets, credentials, API tokens, and private keys should be called from Azure Key Vault rather than stored in code or configuration files.

- For App Service connections to other Azure resources, the connections should be isolated within segmented networks to limit the number of public connections running apps. You should create a dedicated App Service environment through an isolated network tier.

Microsoft Defender for App Service assesses the security posture, provides recommendations, and creates alerts on potential threats and vulnerabilities detected within the App Service environment.

> **Note**
>
> More information on Microsoft Defender for App Service can be found at this link: `https://learn.microsoft.com/en-us/azure/defender-for-cloud/defender-for-app-service-introduction`.

When monitoring and governing App Service and the code of the applications, you need to maintain a secure architecture. A cybersecurity architect should be familiar with the company's standards and policies and continuously review them for compliance.

Specifying Security Requirements for Containers

Containers have become a critical compute component used for modernizing applications to utilize the resiliency, elasticity, and flexibility benefits of cloud technologies. Containers run on hosts where Microsoft provides security monitoring. As a cybersecurity architect, you need to secure the computing resources and the code within the container registry to avoid vulnerabilities and potential threats. Some of the baseline requirements that you should consider are as follows:

- Since containers are an isolated, lightweight portion of a compute host with a host operating system used to run the application, the API and services for the application, as well as the runtime services, should be hardened and protected.

- You should ensure that access and authorization to the container utilize the principles of least privilege to decrease the attack surface.

- Containers use images of multiple layers, and each layer should be monitored for vulnerabilities.

- Azure Container Registry or Docker Registry should be configured to reduce the threat of attacks. Docker registries should be private for the code store and container images.

- Access to Azure Container Registry should utilize Microsoft Entra authentication for resources and service principal authentication. Container data in transit and at rest should have permissions for privileged user access.

- Azure Key Vault can be used to manage secrets to secure access in the development of applications on container platforms.

Defender for Containers assesses the security posture, provides recommendations, and creates alerts on potential threats and vulnerabilities detected within containers and container registries within your Azure, hybrid, and multi-cloud environments.

> **Note**
>
> More information on Microsoft Defender for Containers can be found at this link: `https://learn.microsoft.com/en-us/azure/defender-for-cloud/defender-for-containers-introduction`.

Specifying Security Requirements for Container Orchestration

Containers are built on a cluster management plane and a control plane. The control plane connects the API server for client connections to the nodes that are running the cluster configuration for the application. Security data and events are monitored through Azure Policy and Defender for Cloud through the Azure Security Benchmark.

Microsoft Defender for Containers monitors containers and container registries for vulnerabilities and threats. *Figure 8.11* shows a diagram of the workflow for monitoring the security posture:

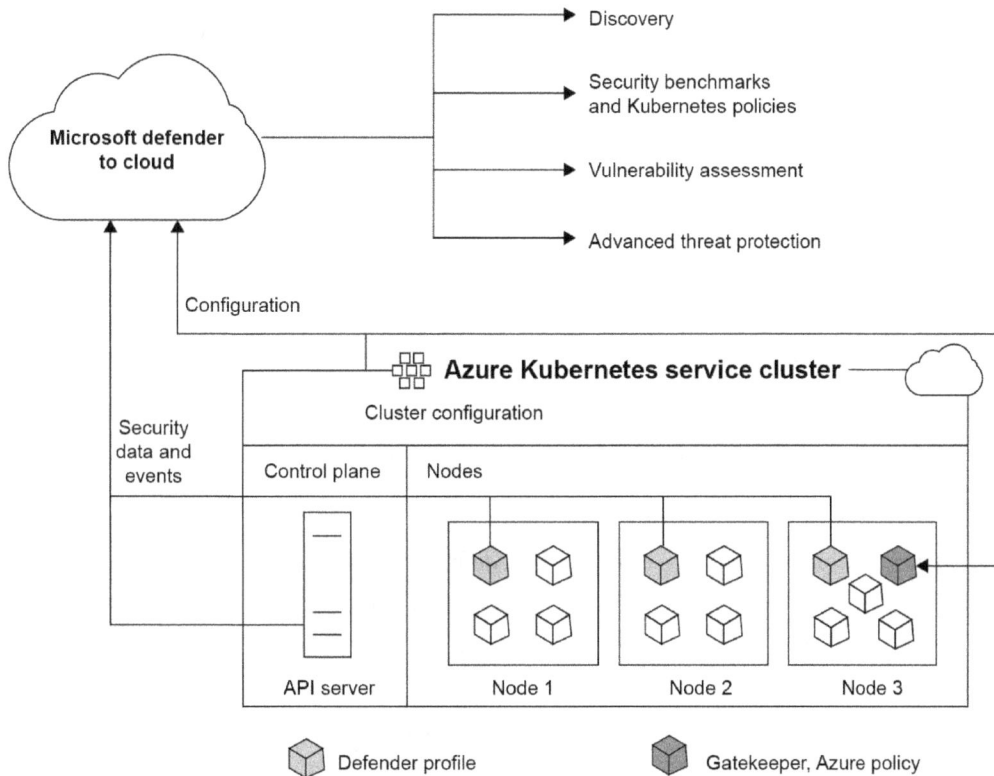

Figure 8.11: Diagram illustrating the architecture of a cloud computing security system, specifically focusing on Microsoft Defender for Cloud integrated with an Azure Kubernetes Service (AKS) cluster

This diagram is useful for understanding how Microsoft's security services integrate with Kubernetes infrastructure to enhance security and compliance and includes the following key sections:

- **Microsoft Defender for Cloud**:

  - **Configuration**: This includes discovery and Kubernetes policies
  - **Security benchmarks**: These feature a vulnerability assessment
  - **Advanced threat protection**: No sub-components are visible

- **Azure Kubernetes Service Cluster**:

  - **Control plane**: This contains an API server connected to a Defender profile, which links to Gatekeeper and Azure Policy
  - **Nodes**: Three nodes labeled *Node 1*, *Node 2*, and *Node 3*, each containing multiple hexagonal icons representing Pods or containers

Evaluating Solutions That Include Azure AI Services Security

What Are Azure AI Services?

Azure AI services are a suite of products and tools that enable organizations to develop applications that integrate AI capabilities at a rapid pace.

There are several services available that include (but are not limited to) the following:

- Azure AI Search (for AI-powered search on desktop and mobile devices)
- Azure OpenAI (allows you to run multiple different AI models for natural language tasks, safely inside your own Azure subscription, where you have control over data, ingress, egress, and security)
- Bot Service (for creating and running AI-powered chatbots)
- Face detection
- Vision (for image recognition)

> **Note**
> For a comprehensive, up-to-date list of available Azure AI services, please check the following link: `https://learn.microsoft.com/en-us/azure/ai-services/what-are-ai-services#available-azure-ai-services`.

Security Considerations

In keeping with the approach taken with other Azure and Microsoft 365 services, they are exposed via both client libraries and APIs, offering flexibility to developers in how they consume these services.

This also means that you can use similar methods to secure them as you would other services.

Azure AI services support several common authentication methods:

- **Authentication headers (for OAuth / access tokens)**: This method involves using OAuth tokens to authenticate API requests. The access tokens are included in the request headers, ensuring secure communication between the client and the service.

  The available authentication headers are as follows:

Header	Description
`Ocp-Apim-Subscription-Key`	Use this header to authenticate with a resource key for a specific service or a multi-service resource key.
`Ocp-Apim-Subscription-Region`	This header is only required when using a multi-service resource key with the Azure AI Translator service. Use this header to specify the resource region.
`Authorization`	Use this header if you are using an access token. The steps to perform a token exchange are detailed in the following sections. The value provided follows this format: `Bearer <TOKEN>`.

Table 8.2 : The headers used for authenticating with Azure services, including resource keys and access tokens

- **Single-service or multi-service resource keys**: These keys function similarly to Azure storage account keys. Each service resource has two keys, allowing for seamless key rotation without interrupting the service. This ensures continuous access while maintaining security.

- **Microsoft Entra ID service principals**: Service principals are identities created for use with applications, hosted services, and automated tools to access Azure resources. They provide a way to authenticate and authorize applications to perform actions on Azure resources.

- **Microsoft Entra ID managed identities**: Managed identities are automatically managed by Azure and provide an identity for applications to use when connecting to resources that support Azure AD authentication. This simplifies the management of credentials and enhances security by eliminating the need to store credentials in code.

Entra would be the most secure option and allows for a more granular RBAC configuration.

Resource keys are less secure and allow less fine-grained access control.

Azure AI services also support Azure virtual networks, allowing you to deploy those services in networks that you control, rather than in a SaaS service that is public.

You can configure network access rules/firewall rules; as a recommended best practice, you should configure your services (under `Resource Management | Networking` in the Azure portal) to only allow access from `Selected Networks and Private Endpoints`, as shown in *Figure 8.12* – this, by default, allows no access until you configure the access that you require to be allowed.

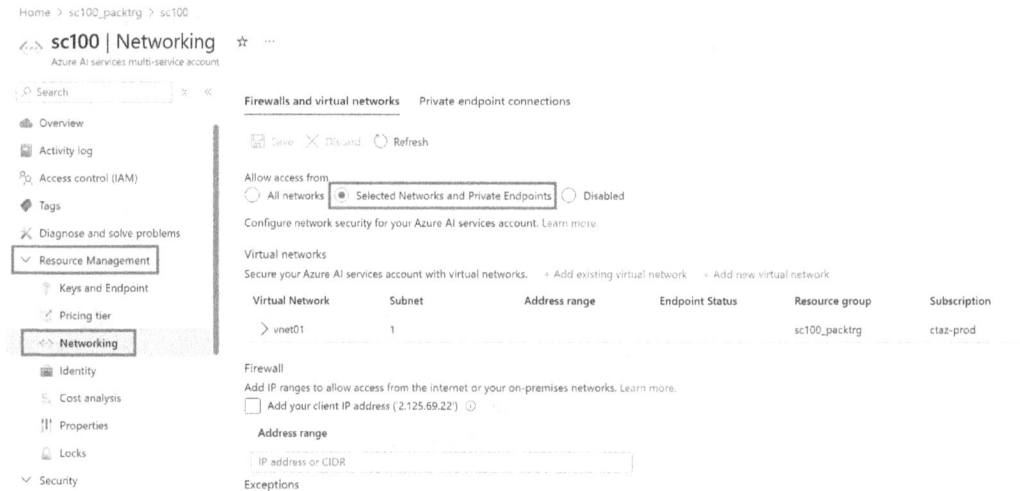

Figure 8.12: Configuring network access for an Azure AI service in the Azure portal

You can, in addition, configure IP network rules to allow specific public internet IPs to access your service, whether they belong to your organization (for example, an on-premises hosted service with a public IP) or not.

As with many other Azure services, you can also configure private endpoints. These enable private communication between two Azure services over the Microsoft backbone, rather than through public interfaces.

A subset of Azure AI services (Azure OpenAI, Azure AI Vision, Content Moderator, custom vision, Face, Document Intelligence, Speech, and QnA Maker) currently also support a DLP control that can allow your AI services to communicate with URLs of your choosing, even if they resolve to IP addresses that would otherwise be blocked by your firewall rules for the service.

Those URLs can be IP address based, such as `https://10.0.0.1/api/v2/$metadata`, as well as **fully qualified domain name** (**FQDN**) based, such as `https://www.myexample.site/api/v2/$metadata`.

This, in combination with the firewall settings, allows a more granular security configuration, such as not allowing access other than to select virtual networks, while still allowing specific URLs for outbound access from the service.

Microsoft also provides an **Azure security baseline** (**ASB**) for the Azure AI services, as they do for most services, which provides guidance on achieving a secure baseline configuration for the service. It is also now known as the **Microsoft cloud security benchmark** (**MCSB**), though both names are still in use at the time of publication of this book.

> **Note**
> You should study this ASB, which is available at this link: `https://learn.microsoft.com/en-us/security/benchmark/azure/baselines/cognitive-services-security-baseline?toc=%2Fazure%2Fai-services%2FTOC.json`.

It also identifies integrations with other security products that we have already discussed extensively in this book, such as Microsoft Defender for Cloud, Microsoft Entra ID Conditional Access, and Microsoft Purview.

> **Note**
> There is extensive support for several regulatory frameworks via built-in Azure Policy definitions for frameworks such as the **Federal Risk and Authorization Management Program** (**FedRAMP**), **Cybersecurity Maturity Model Certification** (**CMMC**), **National Institute of Standards and Technology** (**NIST**), and other popular regulatory/control frameworks. The exam may ask questions that relate to the frameworks covered, although it will not ask about details relating to the controls of the individual frameworks. You can view the full, up-to-date list at this link: `https://learn.microsoft.com/en-us/azure/ai-services/security-controls-policy`.

In combination, these features allow you to design a least-privilege, defense-in-depth architecture to securely deploy Azure AI services while meeting your business and regulatory requirements for cybersecurity.

Case Study – Security Requirements for IaaS, PaaS, and SaaS

Apply what you learned in this chapter by completing the case study on the accompanying online platform. In this case study, you will be given a company scenario and asked to complete several tasks to ensure the company meets the security requirements for IaaS, PaaS, and SaaS.

To access the case study, visit the following link or scan the QR code.

Link to the case study: `https://packt.link/SC100-E2-CaseStudy_Chapter8`

QR code:

Figure 8.13: QR code to access case study for Chapter 8

Summary

This chapter discussed how to design a strategy for securing SaaS, PaaS, and IaaS. It showed you how to build a security baseline for each of these services and security requirements for containers, edge computing, application services, databases, and storage accounts within Azure.

The next chapter will discuss additional security requirements for applications, including prioritizing mitigating threats, standards for onboarding new applications, and security strategies for applications and APIs.

The key takeaways from this chapter are as follows:

- The further along the cloud native journey that you travel, from IaaS to PaaS to SaaS, the less control is available to you to control the infrastructure.

- Regardless of the infrastructure stack, there are core tenets that you are responsible for in each of them that are key to protecting your business and data: identity and access, authentication, data governance/rights management, and secure communications.

- The Defender suite has offerings in all methodologies, be it endpoints, SaaS, PaaS, databases, secrets management, AI, and more. You should assume there will be several questions in the exam that expect you to know what these offerings are, how they are licensed, and how they can be deployed

At a high level, Microsoft Defender licensing can be summarized as follows:

- **Microsoft 365 Defender**

  - **Microsoft 365 Defender**: This suite includes various security services such as Microsoft Defender for Endpoint, Microsoft Defender for Office 365, Microsoft Defender for Identity, and Microsoft Defender for Cloud Apps. The licensing options are as follows:

    - **Microsoft 365 E5 or A5**: This includes full access to Microsoft 365 Defender features

    - **Microsoft 365 E3 with the Microsoft 365 E5 Security add-on**: This provides enhanced security features

    - **Enterprise Mobility + Security (EMS) E5 or A5**: This also includes access to Microsoft 365 Defender features

- **Microsoft Defender for Endpoint**:

  - **Plan 1 (P1)**: This offers core endpoint protection capabilities such as next-generation anti-malware, attack surface reduction, and an endpoint firewall.

  - **Plan 2 (P2)**: This includes all P1 features plus advanced threat protection, **endpoint detection and response (EDR)**, and automated investigation and remediation

- **Microsoft Defender for Cloud**: This provides security management and threat protection across Azure, hybrid, and multi-cloud environments:

  - **Licensing**: Available as a standalone service or included in various Azure subscription plans

- **Microsoft Defender for Office 365**:

  - **Plan 1**: This protects against phishing, malware, and other threats in email and collaboration tools

  - **Plan 2**: Includes all Plan 1 features plus advanced threat protection, automated investigation, and response

- **Microsoft Defender for Identity**:

  - **Licensing**: Available as part of Microsoft 365 E5, A5, or as a standalone service, it helps protect on-premises Active Directory from advanced threats

- **Microsoft Defender for Cloud Apps**:

  - **Licensing**: A user-based subscription service, available as a standalone product or included in various Microsoft 365 and EMS plans

Exam Readiness Drill – Chapter Review Section

Apart from mastering key concepts, strong test-taking skills under time pressure are essential for acing your certification exam. That's why developing these abilities early in your learning journey is critical.

Exam readiness drills, using the free online practice resources provided with this book, help you progressively improve your time management and test-taking skills while reinforcing the key concepts you've learned.

How to Get Started

1. Open the link or scan the QR code at the bottom of this page.
2. If you have unlocked the practice resources already, log in to your registered account. If you haven't, follow the instructions in *Chapter 11* and come back to this page.
3. Once you have logged in, click the **START** button to start a quiz.

We recommend attempting a quiz multiple times till you're able to answer most of the questions correctly and well within the time limit.

You can use the following practice template to help you plan your attempts:

Working On Accuracy		
Attempt	**Target**	**Time Limit**
Attempt 1	40% or more	Till the timer runs out
Attempt 2	60% or more	Till the timer runs out
Attempt 3	75% or more	Till the timer runs out
Working On Timing		
Attempt 4	75% or more	1 minute before time limit
Attempt 5	75% or more	2 minutes before time limit
Attempt 6	75% or more	3 minutes before time limit

The above drill is just an example. Design your drills based on your own goals and make the most of the online quizzes accompanying this book.

First time accessing the online resources? 🔒

You'll need to unlock them through a one-time process. **Head to** *Chapter 11* **for instructions.**

Open Quiz

https://packt.link/SC100_CH08

Or scan this QR code →

Specify Security Requirements for Applications

The previous chapter discussed how to design a strategy for securing SaaS, PaaS, and IaaS infrastructures. This included building a security baseline for each of these services as well as security requirements for the containers, edge computing, application services, databases, and storage accounts within Azure. This chapter will discuss security requirements for applications, including prioritizing mitigating threats, standards for onboarding new applications, and security strategies for applications and APIs.

Understanding the necessity of these security requirements is crucial for safeguarding applications against potential threats and vulnerabilities. By implementing these measures, organizations can protect user identities and ensure the confidentiality and integrity of their data.

This chapter covers the exam domain of **Design security solutions for applications and data**. You are going to cover the following main topics:

- Specifying priorities for mitigating threats to applications
- Specifying a security standard for onboarding a new application
- Designing a security solution for API management
- Specifying a security strategy for applications and APIs
- Case study – security requirements for applications
- Knowledge base questions

Specifying Priorities for Mitigating Threats to Applications

An application is a type of software or tool designed to help you perform specific tasks on a computer or other digital devices. Whether installed on your computer or accessed through a browser, these applications are crucial for conducting business, both personally and professionally. Common tools such as Word, PowerPoint, SharePoint, Teams, and Excel are applications you use daily. These applications use our identities for access, authentication, and authorization to access data that may or may not be confidential or sensitive.

Because of the importance of protecting identity and data within the company, protecting against threats to your applications should be prioritized to identify the risks and mitigate the vulnerabilities and threats. The following sections will provide some guidance for protecting applications from threats and vulnerabilities.

Identity and Secret Handling and Use

Identity refers to the credentials and attributes that uniquely identify a user or an entity within a system. This includes usernames, passwords, and other authentication methods that verify the identity of a user or application.

Secrets are sensitive information that applications use to authenticate and communicate securely. This includes API keys, passwords, certificates, and other confidential data that should be protected from unauthorized access.

The further that you move away from the physical infrastructure and compute resources, the more important it becomes to secure your identity and secrets.

Identity and secrets are combined in this context as the primary ways that users can gain access to an application and how that application can gain access to other applications or databases. They are equally important, and improper handling could cause an identity breach.

To protect against leaked credentials and secrets in our applications, user accounts and secrets should not be exposed within them.

Security operations should be implemented just in time and with just enough access through zero-trust verification methods. By granting access only when it is needed and only to the extent necessary, the risk of unauthorized access is significantly reduced. This approach ensures that users and applications do not have perpetual access to sensitive resources, which minimizes the potential attack surface. If an attacker gains access to a user's credentials, the damage they can cause is limited because access is restricted to specific tasks and timeframes.

If standard security practices have these procedures in place, exposures to vulnerabilities in an application's development and use can be reduced. To further enhance security, Azure Entra ID Conditional Access policies should be employed to validate and verify user and device compliance, thereby mitigating vulnerabilities and threats from brute-force identity attacks.

Additionally, resource secrets should be managed in Azure Key Vault to avoid exposing them in application code.

These secrets should be regularly rotated to mitigate threats created through over-sharing resource secrets within application development and production changes.

Next up is segmentation and configuration.

Segmentation and Configuration

Segmenting virtual networks and databases that house critical information is essential within the security policies for application development. Properly controlling access ensures protection against misconfigurations in applications that could lead to data breaches. While avoiding configuration errors is ideal, minimizing damage when they occur becomes paramount. Some of these misconfigurations may surface during code testing, which you will explore in the next section.

Static and Dynamic Testing

Static and dynamic testing can help identify those configuration errors that were discussed in the previous section. Testing provides peace of mind before an application goes into production. Company procedures need to be in place within a DevSecOps deployment process for this testing. This is an important step to prevent API and code-based vulnerabilities, such as cross-site scripting and injection attacks.

Static application security testing (SAST) is the process of having someone review the code before production. The review of this code is an "eye test" of the written code and not the code in action. Therefore, it may not identify possible vulnerabilities that are only visible when the application is running.

Dynamic application security testing (DAST) is used to test when an application is running. It has the potential to find more vulnerabilities, but since the code is in production, these vulnerabilities are also open for exploitation from an attacker, so monitoring the application during testing is important.

The next section is going to discuss how we can handle data that may be accessed within the application.

Data Handling and Access

The previous topics all relate to protecting the critical asset known as data. Within most applications, there is a data source on the backend in the form of a storage account or a database. You will learn more about securing these data sources in *Chapter 10, Design a Strategy for Securing Data*. This section will focus on a few strategies to consider when it comes to data being accessed within an application.

Data includes personal information such as internal and external user identities, financial information, health information, customer information, and business information, to name a few. A key point to how data is handled and accessed is that an organization must know and understand the data that they are storing. Once they know this data, they must properly label and classify it. Having data properly classified and labelled allows the security team to apply policies that can be used to prevent data leakage through mishandling of the data or giving access to data that users and applications are not authorized to view. Data that you understand as classified or sensitive can be protected with segmentation from public data. Data within databases that are sensitive, such as social security numbers, can also be masked to prevent the data from being viewed by unauthorized users and applications.

The next section will discuss how to address security posture management and workload protection to help mitigate threats and vulnerabilities.

Security Posture Management and Workload Protection

The key to any successful security posture is how applications are monitored and managed for potential anomalous activities and threats. Policies should be in place for all application development so that proper logging takes place and that security operations systems have access to these logs and are reviewed through a proper SIEM/SOAR solution.

Security operations professionals are responsible for having the controls and policies in place to govern the secure development of applications. Having these policies, controls, and information barriers in place will maintain a strong security posture throughout the organization and allow security operations to at least sleep occasionally.

Microsoft Defender for Cloud with the Defender for App Services plan turned on can monitor for potential threats and vulnerabilities while assessing the security posture for improvements that can be made for security controls within Azure. *Chapter 6, Evaluate the Security Posture and Recommending Technical Strategies to Manage Risk*, provided details about Microsoft Defender for Cloud and managing the security posture and protection for workloads for apps and infrastructure. *Figure 9.1* shows the various services, along with application services, databases, storage, and other workloads:

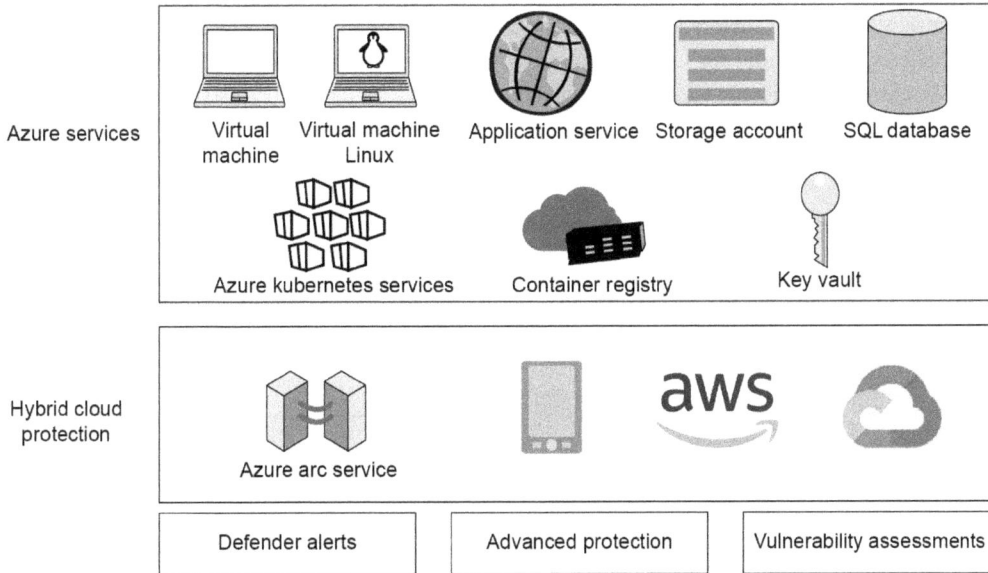

Figure 9.1: Microsoft Defender for Cloud workload protection plans

Microsoft Defender for Cloud Apps within Microsoft 365 Defender can discover and monitor the use of cloud applications. By implementing policies, it helps reduce the use of unsanctioned applications and limits shadow IT.

Microsoft Defender for Cloud Apps is a cloud service with Microsoft 365 that provides **cloud access security broker** (**CASB**) services. A CASB is used as a policy enforcement point between the consumers and the providers so that applications adhere to the baseline security requirements of the company. Microsoft Defender for Cloud Apps provides these capabilities for Microsoft, third-party cloud, and registered on-premises applications.

Microsoft Defender for Cloud Apps is a helpful solution to aid in the discovery of applications that are being used within your company for controlling shadow IT within your company. Shadow IT is the use of applications that are not approved by the company for use on the company network or on devices that also access company data. Microsoft Defender for Cloud Apps identifies all applications accessed by managed users and devices. These applications are then reported on the discovery dashboard. By knowing which applications are being used, we can plan for authorized applications and block unauthorized ones.

For companies that have users in office locations with a firewall or a device that can log network traffic, Microsoft Defender for Cloud Apps' discovery capabilities allow you to connect these logs. Here, Microsoft Defender for Cloud Apps will create a discovery report that lists the applications that are being accessed. The workflow and architecture of Microsoft Defender for Cloud Apps are shown in *Figure 9.2*:

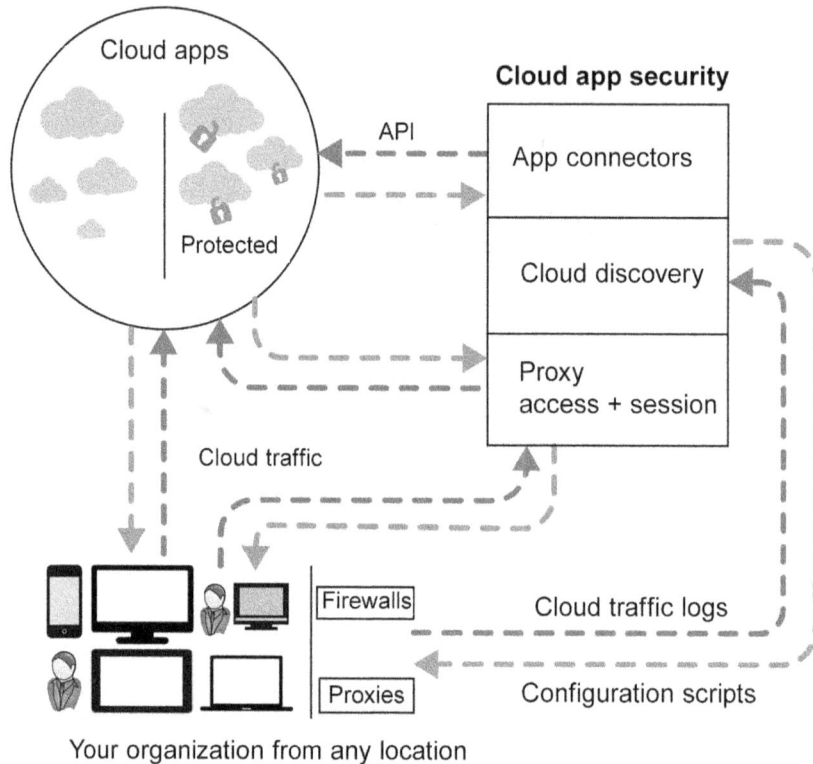

Figure 9.2: Microsoft Defender for Cloud Apps protection workflow

As you continue through this chapter, you will learn about additional requirements for securing various PaaS resources. The next section will discuss security requirements for IoT workloads.

Specifying a Security Standard for Onboarding a New Application

In addition to utilizing tools such as Microsoft Defender for Cloud and Microsoft Defender for Cloud Apps to mitigate threats for applications that are active, you should also consider the full life cycle of an application. This includes the planning, testing, and onboarding of the application. Once that application has been onboarded, you need to monitor, manage, and mitigate that application to protect against threats and vulnerabilities.

To guide you through the application onboarding and the continuous monitoring and compliance life cycle, Microsoft has the **Cloud Adoption Framework** (**CAF**) for Azure. The CAF provides guidance for developing a strategy, building a plan, getting ready to execute that plan, and then adopting the plan by migrating or modernizing workloads to Azure. Once the adoption is complete, the operations and management of the workloads continue the life cycle and drive additional adoption of more workloads. This is a cyclical process for continuous adoption, improvement, and expansion to more modern cloud technologies.

> **Note**
>
> More information on the CAF can be found at this link: `https://learn.microsoft.com/en-us/azure/cloud-adoption-framework/overview`.

The CAF is primarily meant for the migration and modernization of existing applications and data workloads to Azure. For application development, the CAF can still be a valuable framework, but a **development operations** (**DevOps**) approach should also be used for proper testing and approval workflows before an application is onboarded to production. Throughout this process, the security and compliance of the application code, APIs, and application access to data should be identified and should follow company standards, security controls, and compliance requirements. Before embarking on the development process, you should plan the identity strategy and registration of the application within the cloud.

For identity, access to your application can be provided with various authentication techniques:

- Azure Entra ID users are created directly within the Azure Entra ID tenant. These can be member users as well as guest users with external email addresses.

- Azure Entra ID B2B users are external users where a trust relationship has been created between two Azure Entra ID tenants. These users authenticate using their current tenant credentials. The host tenant can grant access to applications through this trust relationship directly or through entitlements.

- Azure Entra ID B2C are users that authenticate with other identity providers that are configured within Azure Entra ID collaboration settings. These are customer relationships that allow access to applications – for example, by using social media accounts for access to a website.

The type of application or platform that you are developing will allow users to be provisioned for access. Utilizing Enterprise SaaS applications within the Azure Entra ID App Gallery provides industry-recognized applications that may already have integration with Azure Entra ID and the ability to auto-provision users in Azure Entra ID. Web apps created within Azure App Service can be registered directly within Azure Entra ID. **System for Cross-Domain Identity Management** (**SCIM**) can be used to create an auto-provisioning API for these applications for users in Azure Entra ID.

> **Note**
>
> Additional information on SCIM can be found at this link: `https://learn.microsoft.com/en-us/azure/active-directory/fundamentals/sync-scim`.

The development process of integrating security and compliance into onboarding applications is known as DevSecOps. The approach of DevSecOps is to have continuous threat, vulnerability, and risk assessment and validation throughout the development of an application. You may have heard this approach to security as "shift left." Shift left moves the security operations from the post-deployment of an application and shifts those operations to the planning and development stages to have security within the overall design of the application.

Applications should go through rigorous testing before going to production, and once in production, should be monitored and improved when a security vulnerability is identified. *Figure 9.3* shows a modified CAF utilizing a DevSecOps approach:

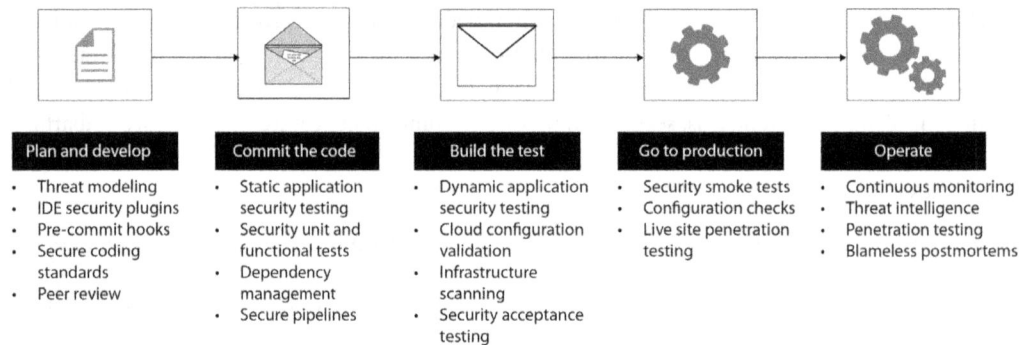

Plan and develop	Commit the code	Build the test	Go to production	Operate
• Threat modeling • IDE security plugins • Pre-commit hooks • Secure coding standards • Peer review	• Static application security testing • Security unit and functional tests • Dependency management • Secure pipelines	• Dynamic application security testing • Cloud configuration validation • Infrastructure scanning • Security acceptance testing	• Security smoke tests • Configuration checks • Live site penetration testing	• Continuous monitoring • Threat intelligence • Penetration testing • Blameless postmortems

Figure 9.3: DevSecOps and the CAF

As previously stated, DevSecOps creates a shift in the approach to security beyond typical operations. Implementing a strategy for onboarding applications with this process requires you to have insights into potential threats at the initial planning and development stages of application creation and then testing throughout the stages of committing the code, building and testing, going to production, and then full operations.

Embracing the shift-left approach involves identifying risks and modeling threats early in the application development process. Establishing a baseline security standard during planning and development enables continuous compliance monitoring from development through testing to production onboarding. If your company lacks an existing application security baseline, consider following the Azure security baseline for comprehensive guidance.

Throughout the DevSecOps process, you should implement a continuous feedback loop to respond to actionable tasks that you may find during testing processes. *Figure 9.4* shows this feedback loop when testing to implement improvements when vulnerabilities are found:

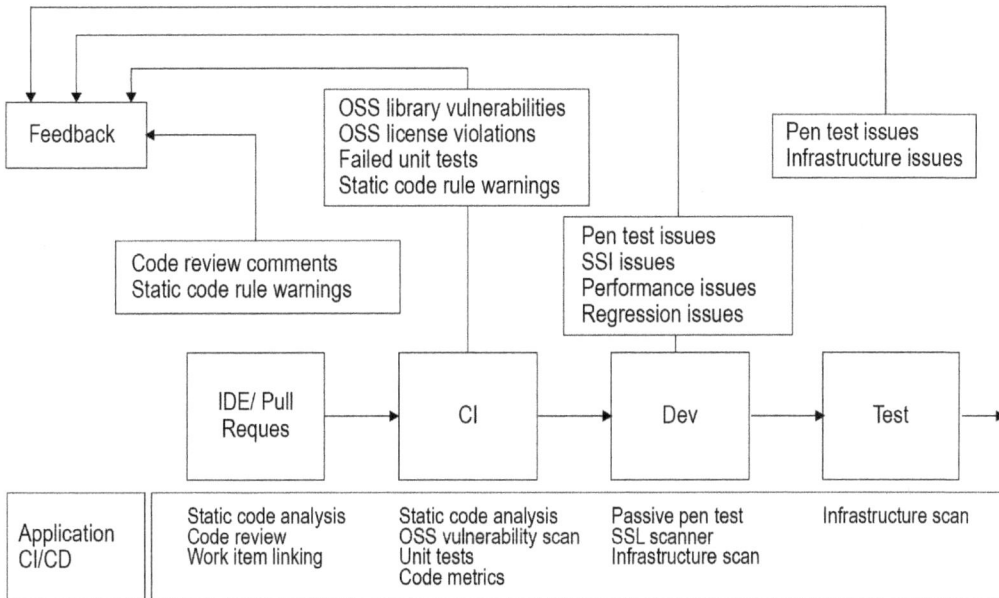

Figure 9.4: Continuous feedback loop

When testing is completed at each stage of the DevSecOps process, the feedback you receive should be properly reviewed and implemented into the application after you've completed a proper risk analysis with the application developers and stakeholders.

> **Note**
>
> For more information on the DevSecOps life cycle process for application development, please see this link: `https://learn.microsoft.com/en-us/azure/cloud-adoption-framework/secure/devsecops-controls`.

DevSecOps is a continuous loop that addresses security from the plan to operations. Cybersecurity architects should be involved throughout this process to recommend controls to maintain the security baseline that you have adopted for your company. Developers and security stakeholders should work collaboratively to prevent applications from being moved to production with potential security vulnerabilities. Once the application is onboarded to production, tools should be put in place, such as Microsoft Defender and Microsoft Sentinel, that continuously monitor and recommend improvements to the application and application infrastructure. *Figure 9.5* shows this continuous DevSecOps life cycle:

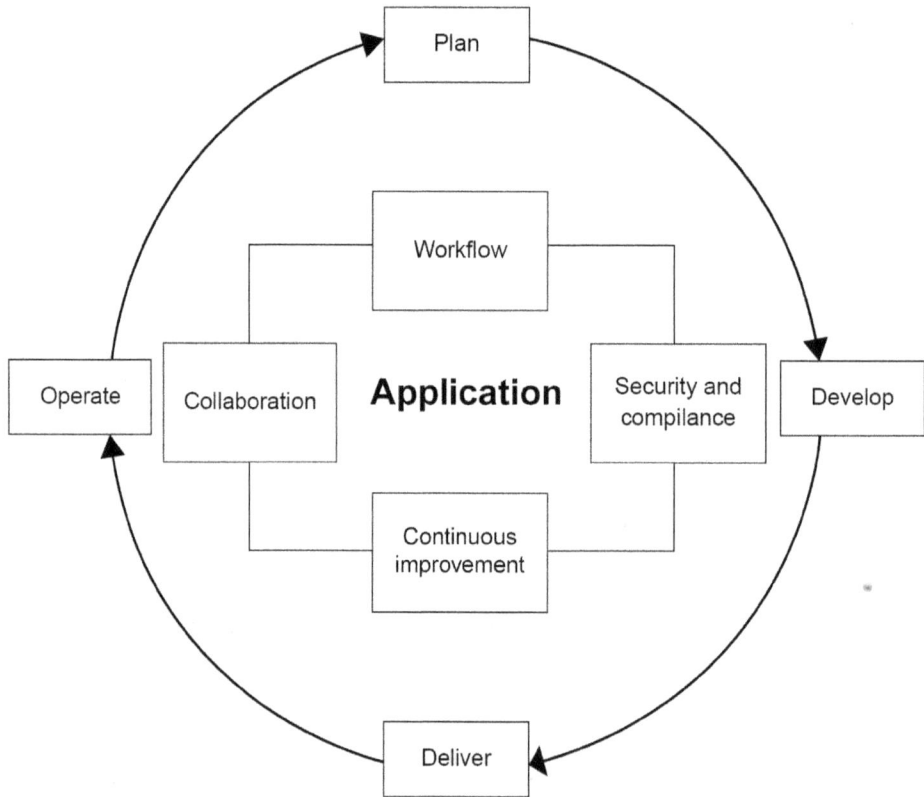

Figure 9.5: Continuous life cycle of DevSecOps

Some application security standards and frameworks that you may want to evaluate as you determine how to implement a shift-left DevSecOps approach are as follows:

- Best practices for application registration: `https://docs.microsoft.com/en-us/azure/active-directory/develop/security-best-practices-for-app-registration`

- Threat modeling tool: `https://docs.microsoft.com/en-us/azure/security/develop/threat-modeling-tool`

- OWASP project for application security verification standards: `https://owasp.org/www-project-application-security-verification-standard/`

- NIST secure software development framework: `https://csrc.nist.gov/publications/detail/sp/800-218/final`

- STRIDE: `https://docs.microsoft.com/en-us/azure/security/develop/threat-modeling-tool-threats`

Any of these standards and frameworks, or a combination of them, can significantly help you with your adoption of a DevSecOps approach to application development.

> **Note**
>
> More information on DevSecOps can be found at this link: `https://learn.microsoft.com/en-us/devops/operate/security-in-devops`.

In the next section, you will learn how to specify a security strategy for applications and APIs.

Designing a Security Solution for API Management

API security management is crucial for organizations building API products. APIs serve as the glue connecting applications and systems, often handling sensitive information. To effectively manage API security, consider implementing robust authentication and authorization protocols (such as OAuth 2.0, JSON web tokens, and OpenID Connect), using **web application firewalls** (**WAFs**) to monitor and filter traffic, ensuring data encryption with TLS/SSL, and validating and sanitizing user input to prevent common attacks such as SQL injection and **cross-site scripting** (**XSS**). Additionally, an API gateway can act as a mediator, providing authentication, rate limiting, monitoring, and access control features. Remember that effective API security management is essential for successful API products, especially if you serve clients in regulated industries such as finance, insurance, or healthcare. Building a proactive, security-first approach ensures reliability and quality for your mobile and web applications.

Figure 9.6 shows the connectivity between devices, APIs, and data sources:

Devices ⟹ APIs ⟹ Data sources

Figure 9.6: Devices sending data to data sources through APIs

> **Note**
>
> For more information on API management, here is the Microsoft Learn link: `Azure API Management - Overview and key concepts | Microsoft Learn`.

Applications serve as the backbone for users to perform daily tasks and access data. Whether internally or externally, the combination of applications and the data they handle holds immense business value. For instance, consider a company website running on Azure App Service, connected to an Azure SQL database. This setup provides real-time inventory information for customers browsing products.

The architecture of applications has evolved with cloud hyperscaler technologies. Legacy applications, previously hosted on servers and virtual machines with embedded identity controls, now embrace modern development using cloud identity providers. Migrating these legacy applications to cloud services often follows a staged approach—rehosting them as **infrastructure-as-a-service (IaaS)** virtual machines in Azure. This process is commonly known as **lift and shift**.

While the lift-and-shift approach allows companies to transition from **capital expenditures (CapEx)** to **operational expenditures (OpEx)**, it doesn't fully unlock Azure's capabilities. Securing rehosted applications remains essential. This involves patching operating systems, conducting SAST and DAST, securing network connections, safeguarding secrets and keys, and ensuring API security. The first step after this lift and shift is to begin to replatform applications, which starts the transition to refactoring. In the replatforming process, developers begin to transition certain components of the application and register the application to Azure Entra ID to utilize the identity and access management security features. Registration of applications into Azure Entra ID can leverage and utilize monitoring and protection of your applications with Microsoft Defender for Cloud Apps and compute resources can be protected by Microsoft Defender for Cloud plans.

The next step is to begin to refactor and redevelop the application to be more cloud-native. This stage in the migration and modernization process will decrease the management overhead, refactoring these applications to utilize **platform-as-a-service** (**PaaS**) solutions within Azure. Utilizing PaaS services, such as Azure App Service, shifts the responsibility of protecting the **operating system** (**OS**) and middleware on the compute resources to the cloud provider, Microsoft, and decreases the attack surface on those resources. By eliminating the most vulnerable attack vector at the compute level, the OS, the security profile of the application shifts to emulate that of a SaaS application. Backend databases should also be refactored while utilizing PaaS data sources, such as Azure SQL Database and Cosmos DB. Securing data and databases will be discussed in further detail in *Chapter 10, Design a Strategy for Securing Data*.

Figure 9.7 is a diagram that shows this shift from an on-premises legacy application to a new development application:

Standalone applications or components of larger solutions

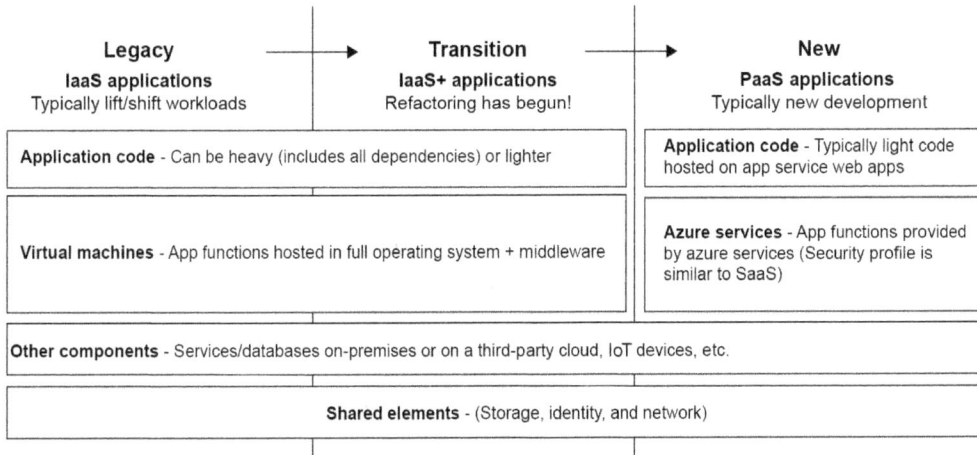

Figure 9.7: Application evolution to PaaS

The shift shown in *Figure 9.7* should be the path that an application developer should be taking toward modernizing an application to be more cloud-native. Cloud-native applications assist security architects and operations by eliminating dependencies on the compute platforms, which, in turn, become security vulnerabilities and potential threats that need to be monitored and patched. APIs can then be designed with this platform independence to be called from any service.

APIs and app functions are provided through secured Azure services. The authorization of APIs should be done while utilizing Azure Key Vault to obscure secrets and keys from being embedded within application code. Compromised secrets can be rotated easily within Azure Key Vault without developer intervention to adjust and test the application code, increasing security and limiting exposure and downtime.

> **Note**
> More information for baseline security for Azure App Service can be found at this link: `https://learn.microsoft.com/en-us/security/benchmark/azure/baselines/app-service-security-baseline`.

The next section will provide you with a scenario where you can apply the concepts that were covered in this chapter.

Case Study – Security Requirements for Applications

Apply what you learned in this chapter by completing the case study on the accompanying online platform. In this case study, you will be given a company scenario and asked to complete several tasks to ensure the company meets the security requirements for applications.

To access the case study, visit the following link or scan the QR code.

Link to the case study: `https://packt.link/SC100-E2-CaseStudy_Chapter9`

QR code:

Figure 9.8: QR code to access case study for Chapter 9

Summary

In this chapter, you learned about the essential security requirements for applications, emphasizing the importance of prioritizing threat mitigation, establishing security standards for onboarding new applications, and designing robust security strategies for applications and APIs. You covered significant examples and concepts such as identity and secrets handling, segmentation and configuration, static and dynamic testing, and data handling and access. This chapter focused on the critical security requirements for applications and the strategies needed to implement them effectively.

In the next chapter, you will discuss designing a strategy for securing data and mitigating threats, including mitigating threats to data, identifying and protecting sensitive data, and data encryption standards.

Exam Readiness Drill – Chapter Review Section

Apart from mastering key concepts, strong test-taking skills under time pressure are essential for acing your certification exam. That's why developing these abilities early in your learning journey is critical.

Exam readiness drills, using the free online practice resources provided with this book, help you progressively improve your time management and test-taking skills while reinforcing the key concepts you've learned.

How to Get Started

1. Open the link or scan the QR code at the bottom of this page.

2. If you have unlocked the practice resources already, log in to your registered account. If you haven't, follow the instructions in *Chapter 11* and come back to this page.

3. Once you have logged in, click the **START** button to start a quiz.

We recommend attempting a quiz multiple times till you're able to answer most of the questions correctly and well within the time limit.

You can use the following practice template to help you plan your attempts:

Working On Accuracy		
Attempt	**Target**	**Time Limit**
Attempt 1	40% or more	Till the timer runs out
Attempt 2	60% or more	Till the timer runs out
Attempt 3	75% or more	Till the timer runs out
Working On Timing		
Attempt 4	75% or more	1 minute before time limit
Attempt 5	75% or more	2 minutes before time limit
Attempt 6	75% or more	3 minutes before time limit

The above drill is just an example. Design your drills based on your own goals and make the most of the online quizzes accompanying this book.

First time accessing the online resources? 🔒

You'll need to unlock them through a one-time process. **Head to** *Chapter 11* **for instructions**.

Open Quiz

`https://packt.link/SC100_CH09`

Or scan this QR code →

10

Design a Strategy for Securing Data

In the previous chapter, you explored the security requirements for applications, focusing on prioritizing the mitigation of threats, establishing standards for onboarding new applications, and developing security strategies for applications and APIs. These accomplishments have laid a strong foundation for our company's data protection strategy in an era where cyber threats are increasingly sophisticated.

In this chapter, we will build on these achievements by exploring how a comprehensive risk management framework, such as the **National Institute of Standards and Technology (NIST) Risk Management Framework (RMF)**, can be applied to assess and mitigate risks to data security. We will also delve into best practices for encryption standards to protect data at rest and in motion, ensuring that sensitive information remains secure even if perimeter defenses are compromised.

This chapter covers the objective domain **Design security solutions for application and data**.

In this chapter, we are going to cover the following main topics:

- Specifying priorities for mitigating threats to data
- Designing a strategy to identify and protect sensitive data
- Specifying an encryption standard for data at rest and in motion
- Case study – designing a strategy to secure data

Specifying Priorities for Mitigating Threats to Data

In addition to people, data is one of the most valuable assets a company possesses. The primary role of a cybersecurity architect is to define and design systems to protect this data. Throughout this book, you have explored different strategies for implementing defense-in-depth and zero-trust methodologies, both aimed at safeguarding data. If you wish to revisit these concepts, refer to *Part 2, Designing a Zero-Trust Strategy and Architecture*.

Consider the financial sector, where data security is of utmost importance. A leading bank faced a critical challenge when it discovered vulnerabilities in its data protection strategy. User identities were being breached, and there was an increased level of privileged access, allowing lateral movement across resources. An identity breach allowed access to data that the compromised identity was still authorized to access. The bank's sensitive financial data was at risk of being exposed, leading to potential exfiltration or deletion. The bank implemented a comprehensive defense strategy, focusing on due diligence and due care in establishing proper controls and monitoring tools. Cybersecurity architects collaborated with security operations to monitor for any signs of account breaches, including auditing for the elevation of privileges. They utilized Microsoft solutions such as Microsoft Entra ID Protection, Privileged Identity Management, Identity Governance, and Conditional Access policies to strengthen their security posture. These tools provided the necessary visibility and control to detect and mitigate threats promptly.

As a result, the bank successfully thwarted an insider attack aimed at breaching a user's identity to gain access to sensitive data. The swift response and robust security measures ensured the protection of the bank's assets and maintained the trust of its customers.

This scenario demonstrates the critical need for vigilance and the effective use of advanced security solutions to protect against data threats in the real world.

Let's discuss how to manage the risk to data.

Managing the Risk to Data

Protecting data is aligned with the ability to manage risks. Identifying potential risks plays a key role in mitigating threats. You should have a framework and process in place to manage and mitigate risks to your company's data. A common framework is the NIST RMF.

> **Note**
> Details on this framework can be found at `https://csrc.nist.gov/Projects/risk-management/about-rmf`.

NIST defines the RMF as a comprehensive, flexible, risk-based approach. The following is from the NIST website:

"The NIST Risk Management Framework (RMF) provides a process that integrates security, privacy, and cyber supply chain risk management activities into the system development life cycle."

Here are the steps that represent the RMF process, as taken from the NIST website:

1. **Prepare**: Essential activities to prepare the organization to manage security and privacy risks
2. **Categorize**: Categorize the system and information processed, stored, and transmitted based on an impact analysis
3. **Select**: Select the set of NIST SP 800-53 controls to protect the system based on risk assessment(s)
4. **Implement**: Implement the controls and document how controls are deployed
5. **Assess**: Assess to determine if the controls are in place, operating as intended, and producing the desired results
6. **Authorize**: A senior official makes a risk-based decision to authorize the system (to operate)
7. **Monitor**: Continuously monitor control implementation and risks to the system

The RMF process starts with preparation. Planning and preparation should be the foundation of any process and framework. In the *Designing a Strategy to Identify and Protect Sensitive Data* section, we will discuss the planning process of identifying the types of data that you have. The process of preparing is the same. Once you have a plan and are prepared to move forward, you must categorize the levels of risk. We discussed threat analysis in the *Threat Analysis* section of *Chapter 2, Build an Overall Security Strategy and Architecture*. We can categorize risks by likelihood and impact, as shown in *Figure 10.1*:

Likelihood		Minor	Moderate	Major
	Very likely	Acceptable risk medium 2	Unacceptable risk high 3	Unacceptable risk extreme 5
	Likely	Acceptable risk low 1	Acceptable risk medium 2	Unacceptable risk high 3
	Unlikely	Acceptable risk low 1	Acceptable risk low 1	Acceptable risk medium 2
	What is the chance that it will happen?	Minor	Moderate	Major

Impact
How Serious is the Risk?

Figure 10.1: Risk assessment categorization

Once risks have been identified and categorized, you must choose whether to implement controls to mitigate those risks. This will include doing a budget analysis and risk exposure calculations. The information in this analysis builds the business case for putting the control in place or accepting the risk. After approval, controls are implemented and assessed for the accuracy of the results. Once these controls have been reviewed for the desired results, risk-based decisions can be made to authorize the continued use of these controls. The final phase of the RMF is continuously monitoring these controls to control risks and manage potential vulnerabilities.

Protecting data and managing risks are determined based on where that data resides. In many cases, on-premises data can be protected with perimeter controls. As we go beyond a company's physical infrastructure, however, data protection becomes more challenging. Cloud technologies and the use of mobile devices create a more open architecture for users but also a wider net to protect data that is being accessed. To protect data access on these devices, you should utilize tools to protect identities, such as Conditional Access policies and **multi-factor authentication (MFA)**. **Mobile device management (MDM)** and **mobile application management (MAM)** with products such as Microsoft Intune can put configuration and compliance policies in place to protect sensitive company data on multiple devices used by company personnel.

In the next section, you will learn more about how to mitigate and protect against a threat that is currently very common: ransomware.

Ransomware Protection and Recovery

Ransomware continues to be a primary method for attackers to harm a company's data. Attackers use ransomware to hold data hostage and extort money from the company for the attacker's financial gain. When a company is affected by a ransomware attack, the attacker gains access to company data and encrypts the data on the company's storage devices or accounts. The attacker then requests financial payment, usually in cryptocurrency, for the attacker to provide the company with the keys required to decrypt its data.

Companies should have a plan in place to protect against these types of attacks, as well as a plan in place to recover from an attack to prevent financial and reputational damage. The best way to plan and recover from a ransomware attack is to prevent such an attack in the first place. A ransomware attack is only successful if the attacker can gain access to all copies of a company's data. Therefore, if proper data backup and recovery methods are architected in the data infrastructure, the ability to recover from an attack becomes more likely.

For on-premises data infrastructures on network storage appliances or databases, you can migrate to cloud platforms that have encryption at rest by default. If you are unable to immediately migrate to a cloud provider, you can use backup and recovery tools such as Azure Backup. Services such as this will back up on-premises data on a schedule and allow low-cost recovery of data. Azure Backup and Azure Site Recovery have low **recovery time objectives (RTOs)** and **recovery point objectives (RPOs)**. RTOs and RPOs measure the time to recovery and the potential amount of data lost, respectively.

For databases, migrating to data platforms that are architected with resilience and recovery capabilities can protect against system and network outages. To protect against ransomware exploitation, it is essential to implement encryption, backups, **data loss prevention** (**DLP**), and zero-trust principles. Azure SQL Database has built-in backup and recovery capabilities with low RTOs and RPOs to recover quickly and limit data loss. Azure storage accounts are architected with encryption at rest by default. Segmenting these accounts to protect sensitive data from public access will make unauthorized access more difficult.

Utilizing due diligence for privileged user accounts will also mitigate these threats. Reviewing user permissions with access reviews that have just-in-time and just-enough access to resources can mitigate potential identity-based attacks that could lead to an attacker executing ransomware or other attacks. *Figure 10.3* shows the life cycle of maintaining identity governance:

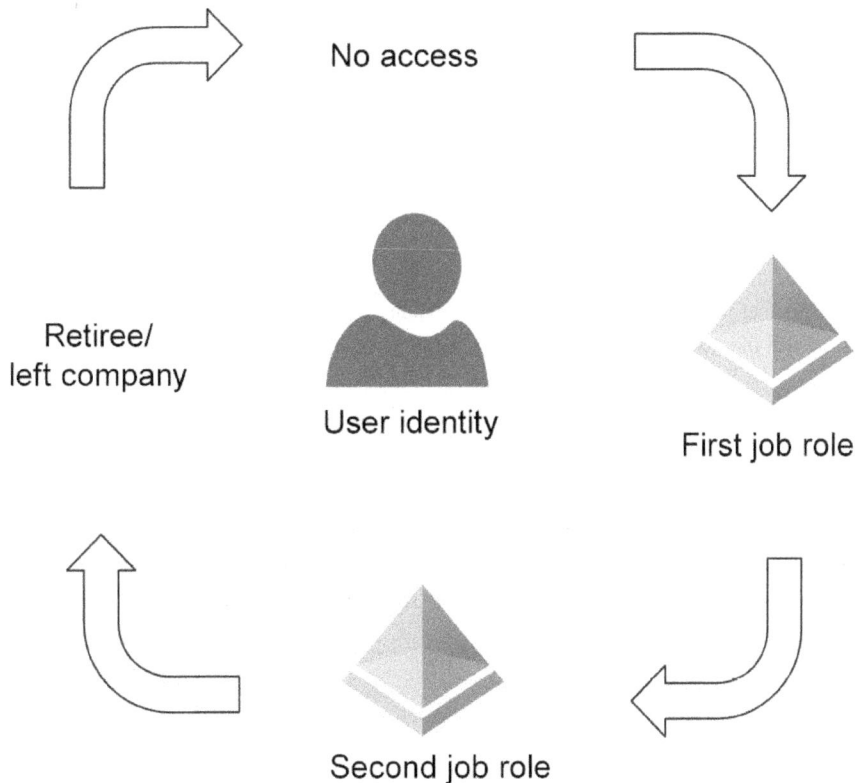

Figure 10.2: Identity governance life cycle

The key phases of protecting against these exploit attacks are to utilize the tools that were previously discussed to do the following:

- Develop a proper recovery plan and have resilience within your data architecture to protect against an attack.

- The recovery plan should have RTO and RPO goals that limit the amount of damage that an attack will cause to the company, both financially and reputationally.

- The identity and access infrastructure should utilize proper identity governance and security to make it harder for an attacker to gain access and, if they do gain access, make it difficult to move laterally through the company infrastructure.

> **Note**
>
> Additional information on Microsoft protection against ransomware can be found at `https://learn.microsoft.com/en-us/azure/security/fundamentals/ransomware-protection`.
>
> Architectural guidance for securing data can be found at `https://learn.microsoft.com/en-us/azure/architecture/data-guide/scenarios/securing-data-solutions`.

To properly protect your data against threats, you must know where your sensitive data is located. The next section will discuss how you can create a strategy to identify and protect sensitive data.

Designing a Strategy to Identify and Protect Sensitive Data

A key component of protecting your data is taking an inventory of your data and assessing the risk to your data. Without knowing what data you have and classifying the level of sensitivity or confidentiality, you cannot properly protect and govern the data. The zero-trust framework for protecting sensitive data is shown in *Figure 10.3*:

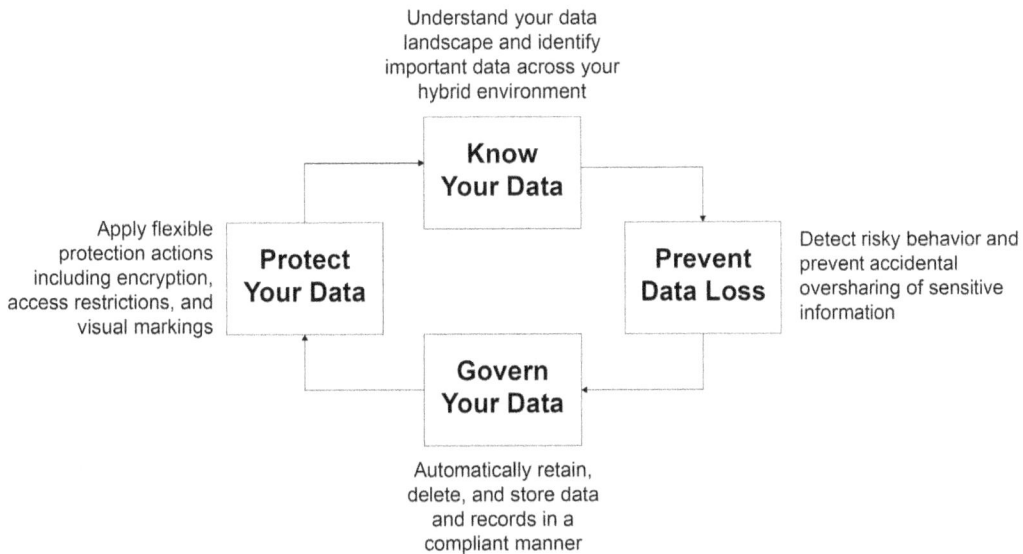

Figure 10.3: Zero-trust framework for data protection

Let's look at each of the steps in this process and see how you can utilize Microsoft tools for each step:

- **Knowing your data** involves taking an inventory of the data that you have and understanding the overall data landscape. This includes identifying the important data that you may have in the cloud and on-premises. Microsoft provides automatic data labeling and classification with the unified labeling client. Data within Azure SQL Database can also be masked to prevent exposure of sensitive data to unauthorized users.

> **Note**
>
> More information on Microsoft sensitivity labeling capabilities can be found at `https://learn.microsoft.com/en-us/microsoft-365/compliance/sensitivity-labels-office-apps?view=o365-worldwide&preserve-view=true#support-for-sensitivity-label-capabilities-in-apps`.

- **Protecting your data** is the utilization of solutions and tools once you understand the data that you have. You may implement different levels of protection, depending on the sensitivity or confidentiality of the data. Data protection solutions within Microsoft for internal and external risks to data can be used, such as Insider Risk Management, Microsoft Defender for Cloud Apps, and Conditional Access policies.

- **Preventing data loss** utilizes solutions to prevent the sharing of sensitive and confidential data. This may include accidental oversharing by internal users. Microsoft Purview Data Loss Prevention can be used to label, protect, and monitor the over-sharing of data.

> **Note**
>
> More information on DLP can be found at `https://learn.microsoft.com/en-us/microsoft-365/compliance/sensitivity-labels-office-apps?view=o365-worldwide&preserve-view=true#support-for-sensitivity-label-capabilities-in-apps`.

- **Governing your data** is the final stage in the data protection process. In this stage, you are determining the life cycle of maintaining the data. Regulations, standards, and business processes may require data to be retained for a specified period to maintain compliance. Microsoft Purview provides the solutions and tools to govern your data as well as protect against and prevent data loss. Information Governance within Microsoft Purview can be used to protect against accidental or malicious deletion of data.

> **Note**
>
> Information on Microsoft Purview can be found here: `https://learn.microsoft.com/en-us/microsoft-365/compliance/?view=o365-worldwide`.

Figure 10.4 shows the process of labeling and classifying data with unified labeling solutions and how this labeling is used throughout the Microsoft Defender and Microsoft Purview solutions for information protection.

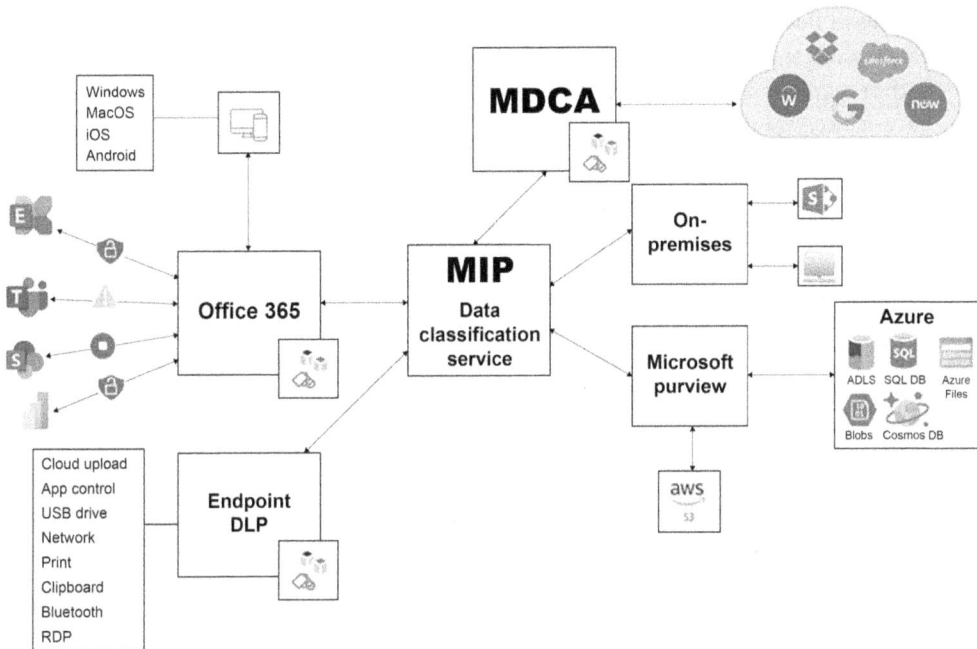

Figure 10.4: Microsoft Information Protection (MIP) workflow

- **Unified labeling solutions**: These solutions provide a consistent and comprehensive approach to classify, label, and protect data across your organization. Labels can be applied automatically based on policies, manually by users, or a combination of both.

- **Labeling and classification**: The process begins with identifying and classifying data based on its sensitivity. Labels such as "Confidential," "Internal," or "Public" can be applied to data to indicate its level of sensitivity and the required protection measures.

- **Microsoft Defender**: Once data is labeled, Microsoft Defender for Cloud Apps uses these labels to enforce protection policies. For example, it can prevent sensitive data from being downloaded, printed, or shared outside the organization. It also helps in identifying potential vulnerabilities and mitigating threats to applications within your company's infrastructure.

- **Microsoft Purview**: This solution governs data throughout its life cycle. It ensures that labeled data is protected against accidental or malicious deletion, and it helps in maintaining compliance with regulations and standards. Microsoft Purview also provides tools for DLP, information governance, and risk management.

- **Zero-trust framework**: This framework is a security model that assumes breaches are inevitable and focuses on verifying every request as though it originates from an open network. It involves strict identity verification, least-privilege access, and continuous monitoring to protect data.

- **Conditional Access policies**: These policies are used to control how and when data can be accessed. They can be configured to require MFA, restrict access based on location, or enforce device compliance. This helps in protecting data from unauthorized access and potential data exposure.

> **Note**
>
> Additional information on the zero-trust framework for securing data can be found at `https://learn.microsoft.com/en-us/security/zero-trust/deploy/data`.

Microsoft Defender for Cloud Apps is a solution that protects data and mitigates threats for applications that are active within your company's application infrastructure. Utilizing the applications that are being accessed within the company can determine potential vulnerabilities that could risk data exposure. Once the applications have been identified, Conditional Access policies can be planned and deployed to prevent data from being downloaded, printed, or shared within the cloud and registered applications. *Figure 10.5* shows the workflow and phases of protection provided within Microsoft Defender for Cloud Apps:

Figure 10.5: Microsoft Defender for Cloud Apps workflow

> **Note**
>
> More information on how data security and privacy can be accomplished with Microsoft Defender for Cloud Apps can be found at `https://learn.microsoft.com/en-us/ defender-cloud-apps/cas-compliance-trust`.

Companies that are utilizing the Microsoft 365 suite of products can also integrate Microsoft Defender for Endpoint with Microsoft Defender for Cloud Apps.

> **Note**
>
> The steps to configure this integration can be found at `https://learn.microsoft. com/en-us/defender-cloud-apps/mde-integration`.

Microsoft Purview follows the framework of knowing your data, protecting your data, preventing data loss, and governing your data. The solutions that are included as part of Microsoft Purview are unified labeling, Information Protection, Data Loss Prevention, and Information Governance.

Microsoft Purview can be used to govern data across SaaS applications, including Microsoft 365 and third-party, on-premises data, and Azure and multi-cloud providers. Separating production and non-production data, along with public and private data from database and storage accounts, should be part of the architecture. The segmented data can then be governed by Microsoft Purview with custom labels and policies for the different data types. *Figure 10.6* shows the segmentation of production and non-production data types within multiple database platforms:

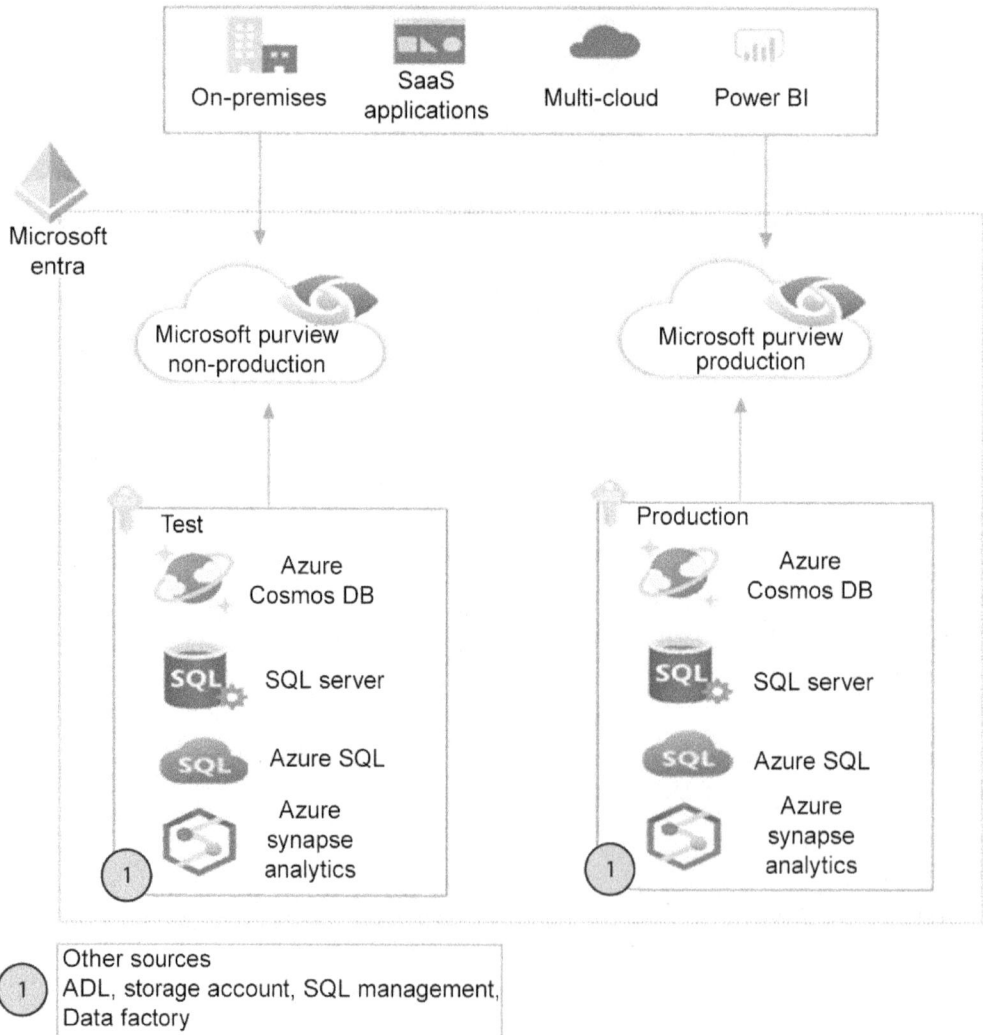

Figure 10.6: Database segmentation for production and non-production

Data in different regions and countries have different requirements in terms of handling and retention. Microsoft Purview can be used to govern the sovereignty of that data within the geographic regions where data is stored and processed on database platforms or storage accounts, as shown in *Figure 10.7*:

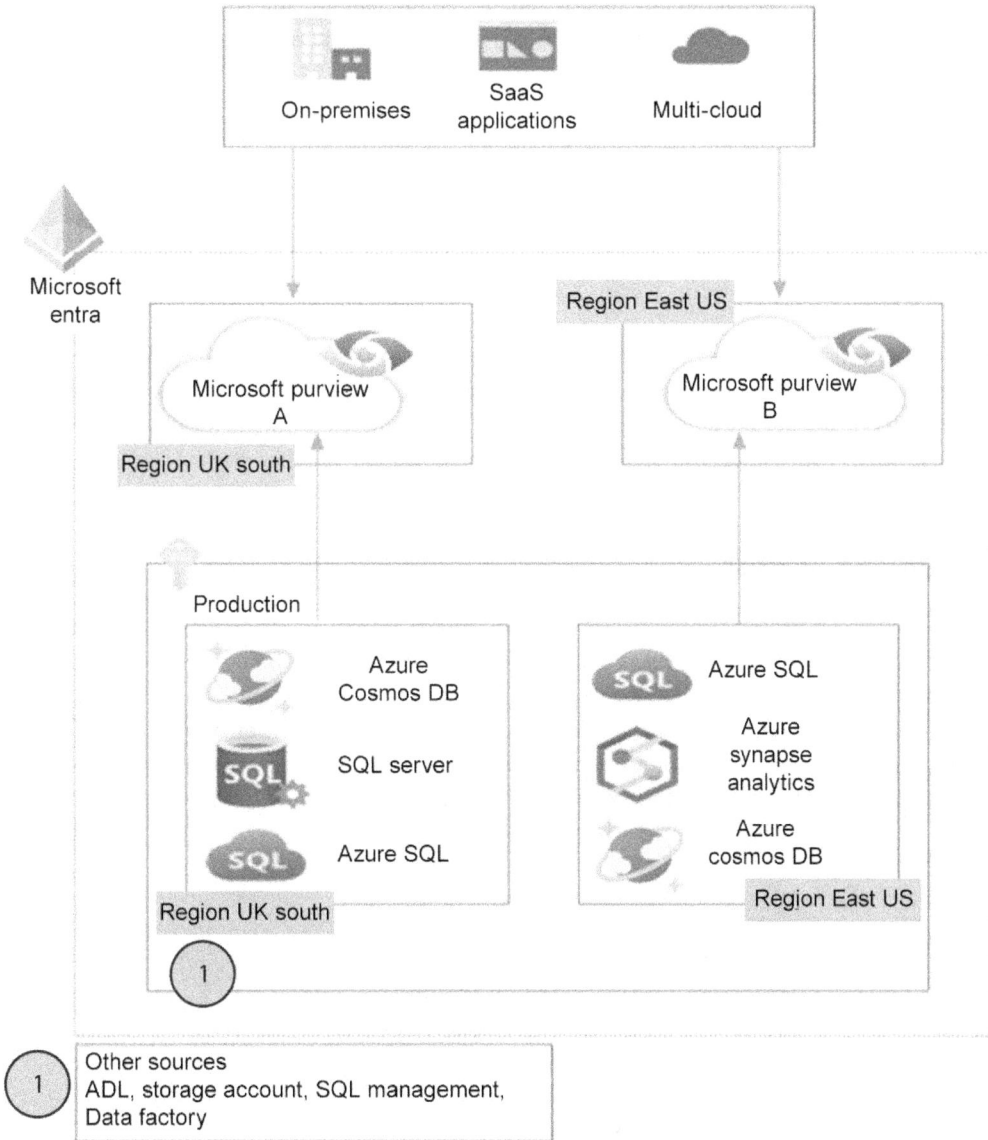

Figure 10.7: Using Microsoft Purview for data sovereignty

Labeling and policies are two ways to prevent data from being exposed, shared, and deleted through Microsoft Purview and other security and governance solutions. However, it is still possible for an internal or external threat to find a vulnerability and access the data. The last line of defense in protecting data is not allowing that data to be exposed and viewed in plain text if it is accessed. Protecting the people who are authorized to access the data and making sure that data is encrypted at rest and in transit provides that final layer of protection.

In the next section, you will learn how to specify an encryption standard for data, both at rest and in motion.

Specifying an Encryption Standard for Data at Rest and in Motion

Whether your data is stored on-premises, in a cloud infrastructure, on SaaS applications, or in a hybrid architecture of all these, you should be encrypting your data. This includes utilizing some form of encryption of data at rest and using secure or encrypted channels for the transmission of data in motion. This section will provide information and solutions that you can use to specify and recommend encryption for your company's data. Let's start with encryption at rest.

Encryption at Rest

Encryption at rest protects data when it is being stored. This is data that resides in a database or a storage account. Having this data encrypted even though it is not being used or transmitted is important. If someone can gain access to a database or storage account and copy this data to a local source, unencrypted data could then be read and exposed. If this data is encrypted at rest, it will be unreadable when copied to another source by an unauthorized user because they will not have the key needed to decipher the data into its plain text form.

The process of how the data can then be accessed by accessing the key and decrypting the data can be seen in *Figure 10.8*:

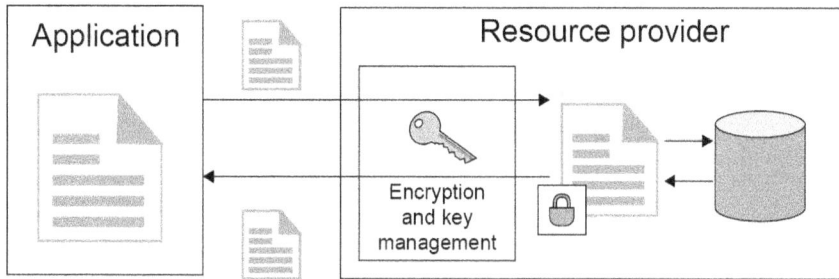

Figure 10.8: Accessing data encrypted at rest

As shown in *Figure 10.9*, the application must authenticate with the key before it can access the data that is stored in its encrypted state.

Encrypting data at rest can be handled with either client-side or server-side encryption. Client-side encryption is outside the services provided by Azure. This encryption is provided by the application, and all the data within Azure is delivered to Azure services already encrypted. Microsoft has no access to the encryption keys, providing a clear separation of duties.

Server-side encryption provides additional models for managing encryption and keys, depending on the level of separation that is required. Key management for server-side encryption may consist of service-managed keys or customer-managed and service-managed keys on customer-controlled hardware. Using service-managed keys on customer-controlled hardware is the most complex option and most Azure services do not support this option.

Azure provides three types of server-side encryption for service-managed keys: **storage service encryption** (**SSE**), **Azure Disk Encryption** (**ADE**), and **transparent data encryption** (**TDE**). The following sections will describe the use of each of them.

Protecting Storage Accounts with SSE

SSE is the encryption service within Azure storage accounts for encrypting data at rest. SSE is turned on by default for all objects and files that are saved within a storage account container or file share. When creating a storage account within Azure, SSE with Microsoft-managed keys is turned on and encryption keys are created for the storage account. These storage account encryption keys can be located and viewed on the **Storage account menu in the Security + networking section as Access keys**. *Figure 10.9* shows this menu:

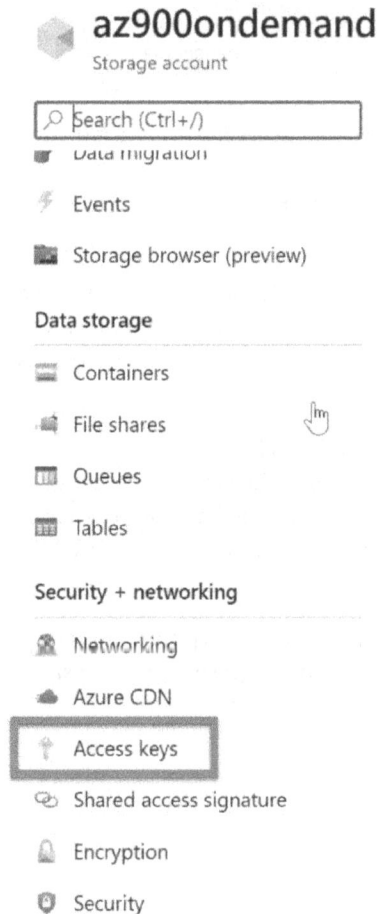

az900ondemand
Storage account

- Search (Ctrl+/)
- Data migration
- Events
- Storage browser (preview)

Data storage

- Containers
- File shares
- Queues
- Tables

Security + networking

- Networking
- Azure CDN
- Access keys
- Shared access signature
- Encryption
- Security

Figure 10.9: Storage account access keys

Within the **Access keys** tile, you can rotate the primary and secondary keys if they have become compromised or are suspected to have been compromised. The **Encryption** tile is where you can change from Microsoft-managed keys to customer-managed keys with Azure Key Vault. We will describe Azure Key Vault later in this section.

> **Note**
>
> For more information on SSE, please visit `https://docs.microsoft.com/en-us/azure/storage/common/storage-service-encryption`.

Next, you will explore ADE for virtual machines.

Protecting Virtual Machines with ADE

ADE is used to encrypt virtual machines and their attached disks. ADE can be used to encrypt virtual machines with Windows and Linux operating systems, and it can also be used to encrypt virtual machine scale sets. ADE for Windows virtual machines utilizes BitLocker for encrypting the virtual machine and attached disks. DM-Crypt is used for Linux virtual machines and attached disks.

> **Note**
>
> For more information on ADE, please visit `https://docs.microsoft.com/en-us/azure/security/fundamentals/azure-disk-encryption-vms-vmss`.

Protecting Databases with TDE

The final type of encryption at rest that will be described is TDE. TDE is encryption at rest for data that is stored within SQL databases, SQL Managed Instance, and Azure Synapse Analytics. These are all platform services within Azure built on Microsoft SQL. TDE is enabled by default on all new SQL databases and SQL Managed Instances but is manually configured on Azure Synapse Analytics.

> **Note**
>
> For more information on TDE, please visit `https://docs.microsoft.com/en-us/azure/azure-sql/database/transparent-data-encryption-tde-overview?tabs=azure-portal`.

Each of these encryption services has keys managed by Microsoft as the default configuration.

Now, let's look at encryption in transit and how it protects data that is being transmitted.

Data Masking

Azure data masking at the database column level is a powerful feature that helps protect sensitive data by obfuscating it for non-privileged users. This feature can be implemented using both static and dynamic data masking techniques. Static data masking replaces sensitive data in a database copy with masked data, ensuring that the original data cannot be retrieved from the masked copy. Dynamic data masking, on the other hand, masks data in real time as queries are executed, without altering the actual data stored in the database.

This allows organizations to limit data exposure while maintaining the usability of their applications. Azure SQL Database, Azure Synapse Analytics, and Azure SQL Managed Instance support granular permissions for dynamic data masking, enabling administrators to control access at the schema, table, and column levels. By using these masking techniques, organizations can enhance their data security and comply with regulatory requirements.

Encryption in Transit

Encryption in transit protects data from being exposed when it is being transmitted over the internet. This is accomplished by using an encrypted and secure transmission channel that uses **Transport Layer Security 1.3 (TLS 1.3)**. TLS 1.3 is the most recent version of the protocol, providing stronger security guarantees and improved performance. When accessing a website, you know that you are using an encrypted channel when it uses HTTPS rather than HTTP. This is extremely important on shopping sites, where you may be required to enter personal information such as addresses and credit card details. Access to this information is protected using your local browser and the data as it is transmitted from your browser to the website, where it cannot be deciphered if it is intercepted.

TLS 1.2+ utilizes a handshake process between the client browser and the server to pass the key and encrypt the data, as shown in *Figure 10.10*:

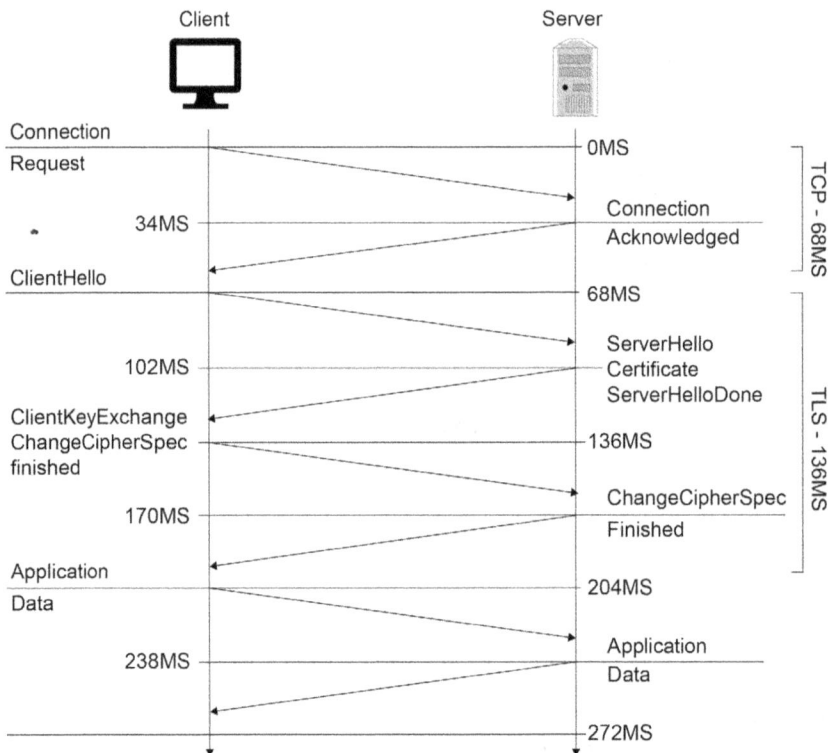

Figure 10.10: TLS 1.2 handshake process

Utilizing encryption in transit with TLS is a common way to encrypt and protect data while in motion. Other ways to protect data in motion include the following:

- **Virtual private network** (**VPN**) connections provide a dedicated connection from one point to another. This could be a location-to-location connection with a site-to-site VPN or a device to a location with a point-to-site VPN. A site-to-site VPN secures the transfer of data by using the **Secure Socket Tunneling Protocol** (**SSTP**) between firewalls. Point-to-site VPNs still utilize the internet for a connection but protect the transmission with an IPsec tunnel.

- Data-link layer encryption in Azure is used for communication between Azure resources and Azure regions. This encryption utilizes the IEEE 802.1AE standard for transmitting data between Azure network connections.

- Azure Storage communication is encrypted when connecting to storage accounts through REST APIs. **Shared access signatures** (**SASs**) are **just-in-time** (**JIT**) links that share files and expire within a defined amount of time. These links utilize TLS encryption with HTTPS links.

- Azure file shares with SMB 3.0 and SMB 3.1.1 are encrypted file shares. Only authenticated and authorized virtual machines and users have access to the file shares.

- **Perfect forward secrecy** (**PFS**) is used to protect client-to-Microsoft cloud service connections with unique 2,048-bit RSA keys that make intercepting and accessing data in transit more difficult.

Once you have determined the type of encryption that will be utilized for your data protection strategy, you need to understand how the keys, secrets, and certificates that are used to encrypt and decrypt the data will be handled.

SSE, TDE, and ADE encryption services can be used with Microsoft-managed keys. However, many standards require some level of separation of duties when it comes to key management. Azure Key Vault can provide this separation of duties. Let's discuss how to use Azure Key Vault to separate key management duties.

Managing Data Encryption Security with Azure Key Vault

If your company has a requirement, whether company or legal, to manage its keys, Azure provides the ability to separate these duties with Azure Key Vault. Azure Key Vault can be used as a centralized location to protect and manage encryption keys, secrets, and certificates.

Figure 10.11 shows where you can configure customer-managed keys within an Azure storage account:

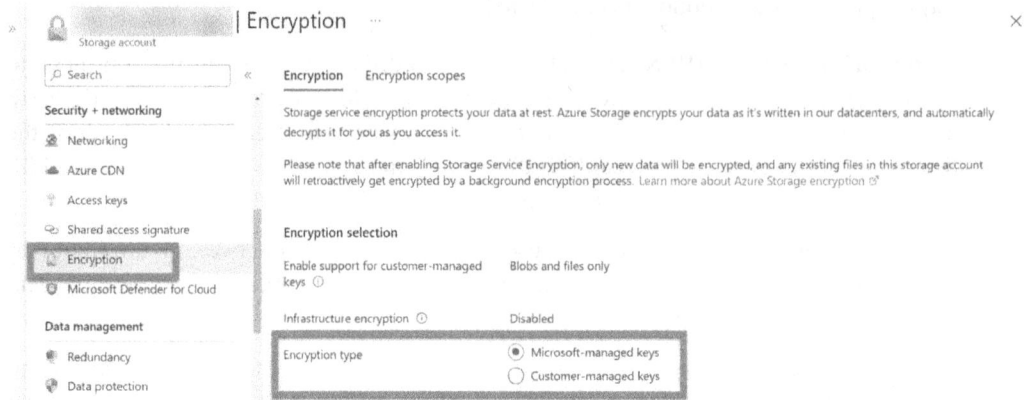

Figure 10.11: Configuring customer-managed keys with Azure Key Vault

Each of the features of security management is listed here:

- **Secrets management**: Securely stores and controls access to tokens, passwords, certificates, API keys, and other secrets.

- **Key management**: Azure Key Vault makes it easy to create and control the encryption keys used to encrypt your data.

- **Certificate management**: Provision, manage, and deploy public and private **Transport Layer Security/Secure Sockets Layer** (**TLS/SSL**) certificates for use with Azure and your internal connected resources. Azure Key Vault can also be configured to automatically renew these certificates before they expire, maintaining secure availability to websites.

- There are two tiers of Azure Key Vault: Standard and Premium. With Premium, you can use Azure Key Vault to synchronize on-premises **hardware security module** (**HSM**) keys and secrets.

Azure Key Vault's capabilities include rotating keys when they may be compromised, and setting automatic rotation to decrease the chance of them being exposed. Utilizing and rotating keys within Azure Key Vault protects you from exposing keys within applications for calls to resources.

Rotating Keys Correctly

Rotating encryption keys is a critical practice to ensure the security of your data. Here is a step-by-step procedure for rotating keys correctly:

1. **Plan the key rotation**: Before starting the key rotation process, plan the rotation carefully. Identify the keys that need to be rotated, the systems and applications that use these keys, and the potential impact on your operations. Ensure that you have a backup of the current keys and a rollback plan in case of any issues.

2. **Generate new keys**: Generate new encryption keys using a secure key management system, such as Azure Key Vault. Ensure that the new keys meet your organization's security requirements and are stored securely.

3. **Update key references**: Update the references to the old keys in your applications and systems to use the new keys. This may involve updating configuration files, environment variables, or database entries. Ensure that the new keys are properly integrated and tested in a non-production environment before deploying to production.

4. **Re-encrypt data (if necessary)**: Depending on your encryption strategy, you may need to re-encrypt your data with the new keys. This can be done gradually to minimize the impact on your operations. Ensure that the re-encryption process is thoroughly tested and monitored.

5. **Update key rotation policies**: Update your key rotation policies to reflect the new keys and the rotation schedule. Ensure that your policies are documented and communicated to all relevant stakeholders.

6. **Monitor and validate**: Monitor the key rotation process to ensure that it is completed successfully. Validate that the new keys are being used correctly and that there are no issues with data access or encryption. Address any issues promptly and update your rollback plan as needed.

7. **Revoke old keys**: Once the new keys are fully integrated and validated, revoke the old keys to prevent their use. Ensure that the old keys are securely deleted and that there are no lingering references to them in your systems.

> **Note**
>
> For more information on Azure Key Vault, please visit `https://docs.microsoft.com/en-us/azure/key-vault/general/overview`.

The next section will discuss more ways to handle secrets and identities to protect data.

Identity and Secret Handling and Use

To protect against leaked credentials and secrets in our applications, user accounts and secrets should not be exposed within the applications. Security operations should be implemented just in time and with just enough access through zero-trust verification methods. If standard security practices have these procedures in place, this will limit data exposure in an application's development and use.

Resource secrets should be managed in Azure Key Vault, which will help you prevent them from being exposed in application code. As stated in the previous section, secrets should be regularly rotated to mitigate against threats created through over-sharing resource secrets within application development and production changes.

Azure Key Vault should be used with a clear strategy for key and secret use and rotation, along with certificate renewals. This will decrease the potential for exposure as well as maintain access to data for authorized users, devices, and resources.

> **Note**
>
> More information on encryption best practices can be found at `https://learn.microsoft.com/en-us/azure/security/fundamentals/data-encryption-best-practices`.

The next section will provide you with a scenario where you can apply the concepts that were covered in this chapter.

Case Study – Designing a Strategy to Secure Data

Apply what you learned in this chapter by completing the case study on the accompanying online platform. In this case study, you will be given a company scenario and asked to complete several tasks to meet the requirements for designing a strategy to secure data.

To access the case study, visit the following link or scan the QR code.

Link to the case study: `https://packt.link/SC100-E2-CaseStudy_Chapter10`

QR code:

Figure 10.12: QR code to access case study for Chapter 10

Summary

In this chapter, you have explored the design of a comprehensive strategy for securing data and mitigating threats. You have covered key topics such as mitigating threats to data, identifying and protecting sensitive data, and establishing data encryption standards. Significant examples included the use of information protection for labeling sensitive data and the application of Microsoft Purview to govern data with labels that can be utilized across Microsoft 365, Azure, and multi-cloud resources.

By going through this chapter, you have gained a detailed understanding of how to implement these strategies and best practices to enhance data protection measures. This knowledge will be crucial in the subsequent chapters, especially in relation to the *SC-100 Microsoft Cybersecurity Architect* exam.

Exam Readiness Drill – Chapter Review Section

Apart from mastering key concepts, strong test-taking skills under time pressure are essential for acing your certification exam. That's why developing these abilities early in your learning journey is critical.

Exam readiness drills, using the free online practice resources provided with this book, help you progressively improve your time management and test-taking skills while reinforcing the key concepts you've learned.

How to Get Started

1. Open the link or scan the QR code at the bottom of this page.
2. If you have unlocked the practice resources already, log in to your registered account. If you haven't, follow the instructions in *Chapter 11* and come back to this page.
3. Once you have logged in, click the **START** button to start a quiz.

We recommend attempting a quiz multiple times till you're able to answer most of the questions correctly and well within the time limit.

You can use the following practice template to help you plan your attempts:

Working On Accuracy		
Attempt	Target	Time Limit
Attempt 1	40% or more	Till the timer runs out
Attempt 2	60% or more	Till the timer runs out
Attempt 3	75% or more	Till the timer runs out
Working On Timing		
Attempt 4	75% or more	1 minute before time limit
Attempt 5	75% or more	2 minutes before time limit
Attempt 6	75% or more	3 minutes before time limit

The above drill is just an example. Design your drills based on your own goals and make the most of the online quizzes accompanying this book.

First time accessing the online resources? 🔒
You'll need to unlock them through a one-time process. **Head to** *Chapter 11* **for instructions.**

Open Quiz	
https://packt.link/SC100_CH10	
Or scan this QR code →	

11
Accessing the Online Practice Resources

Your copy of *Microsoft Cybersecurity Architect Exam Ref SC-100, Second Edition* comes with free online practice resources. Use these to hone your exam readiness even further by attempting practice questions on the companion website. The website is user-friendly and can be accessed from mobile, desktop, and tablet devices. It also includes interactive timers for an exam-like experience.

How to Access These Materials

Here's how you can start accessing these resources depending on your source of purchase.

Purchased from Packt Store (packtpub.com)

If you've bought the book from the Packt store (`packtpub.com`) as an eBook or in print, head to `https://packt.link/SC100e2Unlock`. There, log in using the same Packt account you created or used to purchase the book.

Packt+ Subscription

If you're a *Packt+ subscriber*, you can head over to the same link (`https://packt.link/SC100E2Practice`), log in with your **Packt ID**, and start using the resources. You will have access to them as long as your subscription is active.

If you face any issues accessing your free resources, contact us at `customercare@packt.com`.

Purchased from Amazon and Other Sources

If you've purchased from sources other than the ones mentioned previously (such as **Amazon**), you'll need to unlock the resources first by entering your unique sign-up code provided in this section. **Unlocking takes less than 10 minutes, can be done from any device, and needs to be done only once**. Follow these five easy steps to complete the process:

Step 1

Open the link `https://packt.link/SC100e2Unlock` or scan the following **QR code** (*Figure 11.1*):

Figure 11.1: QR code for the page that lets you unlock this book's free online content

Either of those links will lead to the following page:

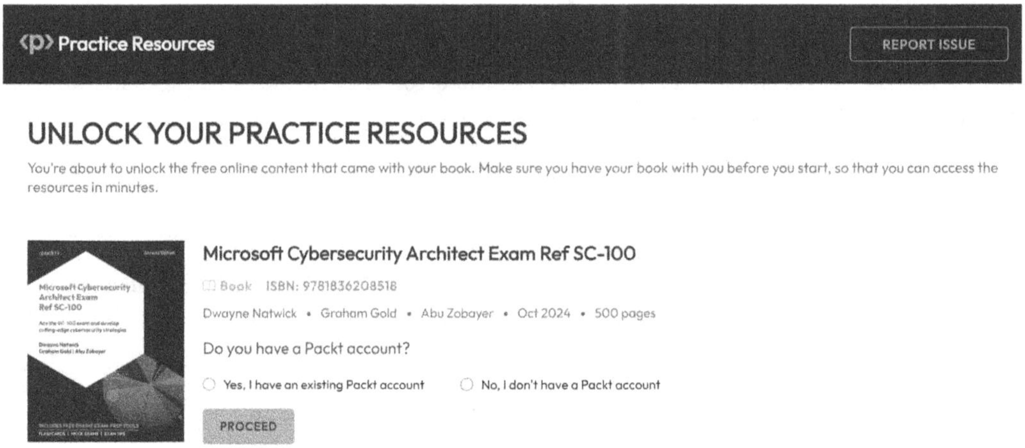

Figure 11.2: Unlock page for the online practice resources

Step 2

If you already have a Packt account, select the option **Yes, I have an existing Packt account. If not, select the option No, I don't have a Packt account**.

If you don't have a Packt account, you'll be prompted to create a new account on the next page. It's free and only takes a minute to create.

Click **Proceed** after selecting one of those options.

Step 3

After you've created your account or logged in to an existing one, you'll be directed to the page shown in *Figure 11.3*.

Make a note of your unique unlock code:

`ZJT8768`

Type in or copy this code into the text box labeled **Enter Unique Code**:

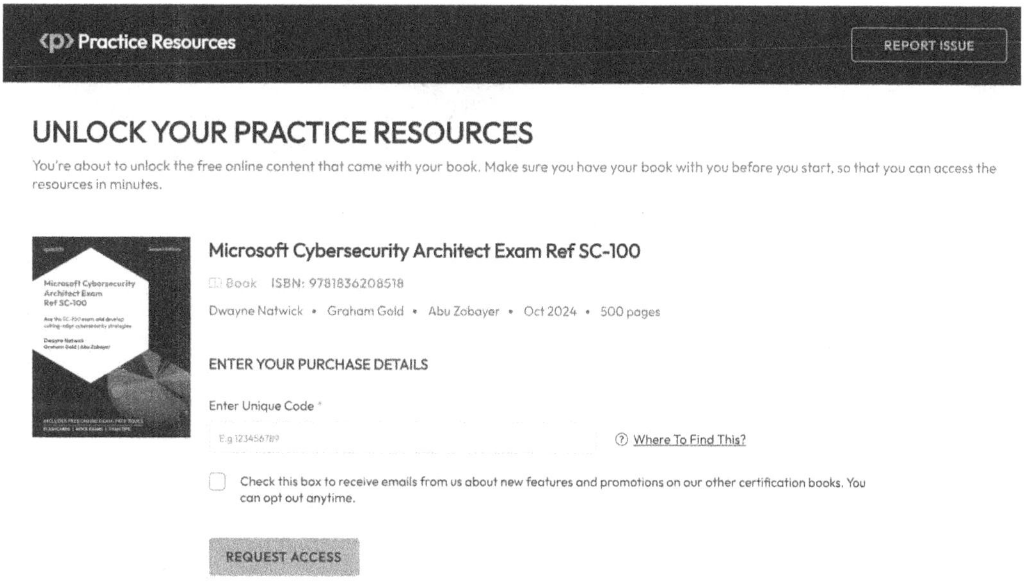

Figure 11.3: Enter your unique sign-up code to unlock the resources

> **Troubleshooting tip**
>
> After creating an account, if your connection drops off or you accidentally close the page, you can reopen the page shown in *Figure 11.2* and select **Yes, I have an existing account**. Then, sign in with the account you had created before you closed the page. You'll be redirected to the screen shown in *Figure 11.3*.

Step 4

> **Note**
>
> You may choose to opt into emails regarding feature updates and offers on our other certification books. We don't spam, and it's easy to opt out at any time.

Click **Request Access.**

Step 5

If the code you entered is correct, you'll see a button that says **OPEN PRACTICE RESOURCES**, as shown in *Figure 11.4*:

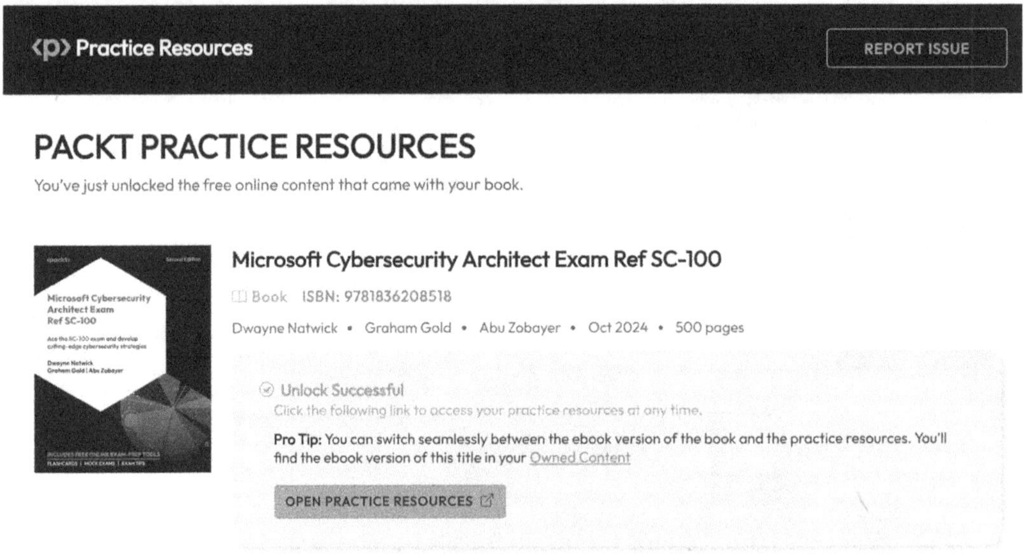
Practice Resources
REPORT ISSUE
PACKT PRACTICE RESOURCES
You've just unlocked the free online content that came with your book.
Microsoft Cybersecurity Architect Exam Ref SC-100
Book ISBN: 9781836208518
Dwayne Natwick • Graham Gold • Abu Zobayer • Oct 2024 • 500 pages
Unlock Successful
Click the following link to access your practice resources at any time.
Pro Tip: You can switch seamlessly between the ebook version of the book and the practice resources. You'll find the ebook version of this title in your Owned Content
OPEN PRACTICE RESOURCES

Figure 11.4: Page that shows up after a successful unlock

Click the **OPEN PRACTICE RESOURCES** link to start using your free online content. You'll be redirected to the dashboard shown in *Figure 11.5*:

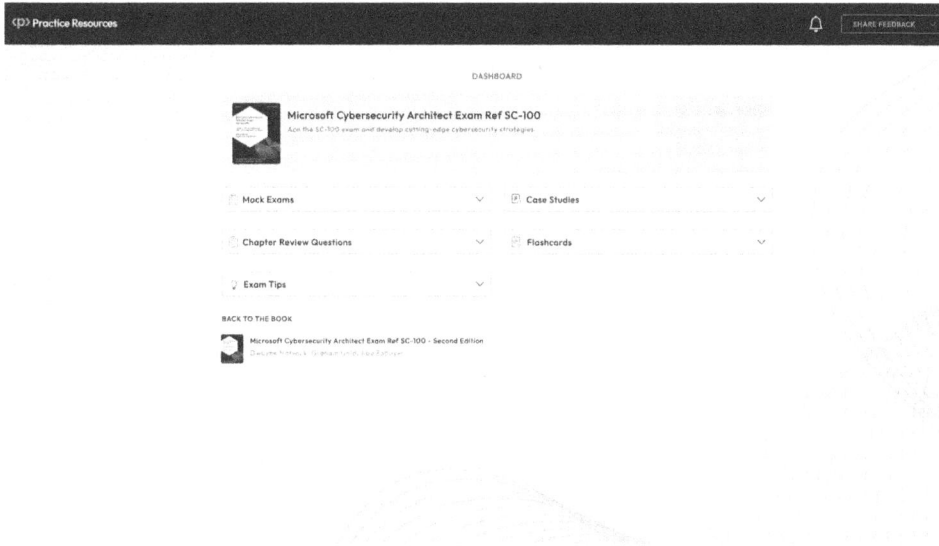

Figure 11.5: Dashboard page for SC-100 practice resources

Bookmark this link

Now that you've unlocked the resources, you can come back to them anytime by visiting `https://packt.link/SC100E2Practice` or scanning the QR code provided in *Figure 11.6*:

Figure 11.6: QR code to bookmark practice resources website

Troubleshooting Tips

If you're facing issues unlocking, here are three things you can do:

- Double-check your unique code. All unique codes in our books are case-sensitive and your code needs to match exactly as it is shown in *Step 3*.

- If that doesn't work, use the **Report Issue** button located at the top-right corner of the page.

- If you're not able to open the unlock page at all, write to `customercare@packt.com` and mention the name of the book.

Share Feedback

If you find any issues with the platform, the book, or any of the practice materials, you can click the **Share Feedback** button from any page and reach out to us. If you have any suggestions for improvement, you can share those as well.

Back to the Book

To make switching between the book and practice resources easy, we've added a link that takes you back to the book (*Figure 11.7*). Click it to open your book in Packt's online reader. Your reading position is synced so you can jump right back to where you left off when you last opened the book.

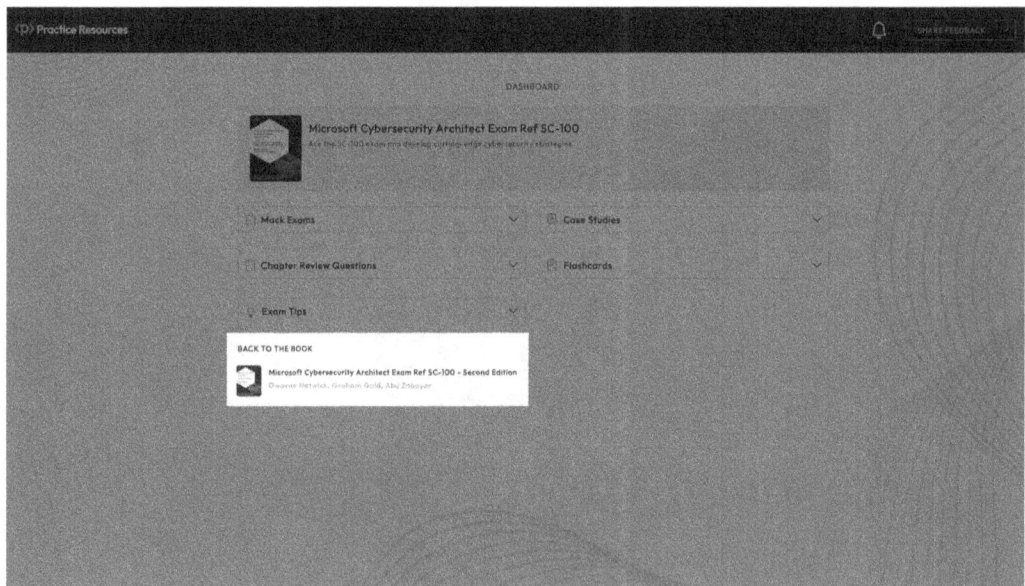

Figure 11.7: Dashboard page for SC-100 practice resources

> **Note**
> Certain elements of the website might change over time and thus may end up looking different from how they are represented in the screenshots of this book.

Index

N

O

‹packt›

Other Books You May Enjoy

If you enjoyed this book, you may be interested in these other books by Packt:

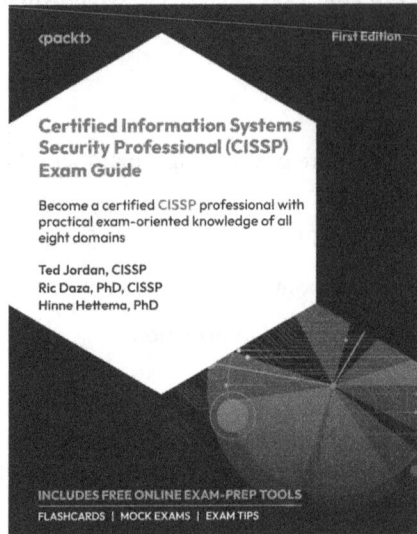

Certified Information Systems Security Professional (CISSP) Exam Guide

Ted Jordan, Ric Daza, and Hinne Hettema

ISBN: 978-1-80056-761-0

- Get to grips with network communications and routing to secure them best
- Understand the difference between encryption and hashing
- Know how and where certificates and digital signatures are used
- Study detailed incident and change management procedures
- Manage user identities and authentication principles tested in the exam
- Familiarize yourself with the CISSP security models covered in the exam
- Discover key personnel and travel policies to keep your staff secure
- Discover how to develop secure software from the start

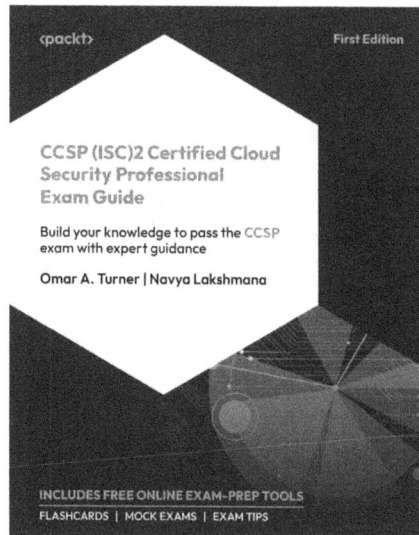

CCSP (ISC)2 Certified Cloud Security Professional Exam Guide

Omar A. Turner and Navya Lakshmana

ISBN: 978-1-83898-766-4

- Gain insights into the scope of the CCSP exam and why it is important for your security career
- Familiarize yourself with core cloud security concepts, architecture, and design principles
- Analyze cloud risks and prepare for worst-case scenarios
- Delve into application security, mastering assurance, validation, and verification
- Explore privacy, legal considerations, and other aspects of the cloud infrastructure
- Understand the exam registration process, along with valuable practice tests and learning tips

Share Your Thoughts

Now you've finished *Microsoft Cybersecurity Architect Exam Ref SC-100, Second Edition*, we'd love to hear your thoughts! Scan the QR code below to go straight to the Amazon review page for this book and share your feedback or leave a review on the site that you purchased it from.

`https://packt.link/r/1836208510`

Your review is important to us and the tech community and will help us make sure we're delivering excellent-quality content.

Download a Free PDF Copy of This Book

Thanks for purchasing this book!

Do you like to read on the go but are unable to carry your print books everywhere?

Is your eBook purchase not compatible with the device of your choice?

Don't worry – now, with every Packt book, you get a DRM-free PDF version of that book at no cost.

Read anywhere, any place, on any device. Search, copy, and paste code from your favorite technical books directly into your application.

The perks don't stop there: you can get exclusive access to discounts, newsletters, and great free content in your inbox daily.

Follow these simple steps to get the benefits:

1. Scan the QR code or visit the link below:

https://packt.link/free-ebook/9781836208518

2. Submit your proof of purchase.
3. That's it! We'll send your free PDF and other benefits to your inbox directly.

www.ingramcontent.com/pod-product-compliance
Lightning Source LLC
Chambersburg PA
CBHW081054220326
41598CB00038B/7097